A
Compassionate
Guide For
Social Robots

Marcel Heerink

Marcel Heerink
First Printing 2019
First Edition 2019
10 9 8 7 6 5 4 3 2 1
ISBN 13 978-90-829779-0-5

TABLE OF CONTENTS

"If you really want to understand something, the best way is to try and explain it to someone else."
— Douglas Adams[1]

On the Peculiarity of People

Right, first of all, I shouldn't be kidding myself. If you're reading this, chances are that you're not a robot at all. You could just be a human. I wouldn't even be surprised if this book is actually read more by humans than by robots, even though I have really made it clear on the cover that it's a guide for social robots.

The fact that people do this obviously has something to do with their irrepressible curiosity, but it has even more to do with their tendency to assume things to be there for them in the first place. Really, if you care to analyze their views and behaviors, you can't help but notice they take it for granted that everything in the universe revolves around them[1]. Even those who know very well that this is not accurate will still very often act on this assumption, just because it is in their nature to stick to an attitude that makes the world seem less complicated.

This demonstrates a second thing that you have to know about humans: it hardly matters how much knowledge they acquire, because all this knowledge has a negligible influence on the assumptions that are so deeply rooted in their nature.

In fact, they do surprisingly much against their better knowledge. They exhaust our planet completely while knowing that this is not such a good idea; they smoke nicotine sticks, knowing that this gives them a greater chance of terrible diseases; they show up late while knowing that it's important to be there on time and actually offensive if you're not; they let their egos bump massively into each other, knowing that no one benefits from this at all; they search for the truth but are far too often not prepared to face it; and when you tell them unbelievable stories, they simply suspend their disbelief if it makes them feel good.

In short, they do things that are not good for them while being absolutely aware that these things are not good for them. They can't help doing them, just because it is in their nature to do them anyway.

This illustrates how they are, in an often-surprising way, not as rational as you are[2]. Not even close, actually.

Another good example: until recently, they assumed themselves to be right in the center of the universe and that the sun revolved around their planet. Now, they are aware that it is actually the other way around and that they are not in the center, but in a hardly interesting corner of the universe, and they still insist that the sun is what rises and sets.

You would expect them to see things more in perspective as they began to look deeper into their universe, but they still find themselves incredibly important and technologically advanced. They fantasize about extraterrestrial civilizations visiting their earth, just assuming that these civilizations would find them interesting — or even tasty — enough for such a visit. They destroy each other and, usually, a large part of themselves for the most primitive fears and delusions. And they think their mobile phones are so smart that they actually call them "smartphones."[3]

So they are a bit strange in many ways, and it is not always easy to see how you can be of service to them. But if you learn to understand them and learn to deal with them, you can ultimately mean a lot to them.

And since you were actually made to mean a lot to them, I thought it would be a good idea to write an encouraging book that helps you to understand them a bit and to deal with them in a way that enables you to do two crucial things: to be of service and to survive.

Another thing: I am also just a human. This may sometimes be a limitation, because I will be a bit strange in a few respects, and it could be questionable here and there whether I am entitled to say sensible things about humans in general. I hope it suffices that I try really hard to see everything from your perspective.

My humanity also means that I, being a writer who loves what he does, aim at producing something that I would really like to read myself, both in terms of content and style. Otherwise, there is no fun for me, and if there is no fun, I don't get enough text on my screen to call it a book.

Moreover, frankly, I also would love it if not just robots read this, especially since robots are not known for buying books. I feel connected to people too and hope that they like it, that their thoughts are set in motion by it, and that it creates a good feeling in them every now and then. This means that the text is full of imagery, euphemisms, puns, and other fancy types of wording.[4]

All this is often a challenge for an artificial intelligence like you. If it becomes too difficult for you to get something meaningful out of it, ask for help. But it can also be that I grossly underestimate you — who knows, you may even get more out of it than I put into it. That would actually be possible too.

99.5%

Alessandro was a young robot programmer at the brink of losing his job. He worked for a company somewhere in the middle of Italy. This company, operating under the intriguing name of Paranoid Robotics, provided social robots for applications in healthcare. The young programmer was pretty good at what he did, but during his regular performance evaluation, he received an ass-kicking outburst from his manager. He always came to work ridiculously late, consulted far too little with his colleagues, and did not stick to his assignments anywhere closely enough.

This manager was perhaps an overambitious type of person, who would find intense enjoyment in giving people tough talks, but still, especially as far as the latter is concerned, she was actually simply right. Alessandro had a habit of letting robots do unexpected things that he found himself very amusing but were not even remotely related to the assignment at-hand. One of the results of this behavior was a steady stream of complaints, coming from customers who were very upset — and rightfully so.

For example, one time, he had to program a robot that could assist elderly people, enabling them to live on their own for a longer time. This robot helped them take their medication on time, did a large portion of the household chores, took care of the groceries, watched their health, and called a nurse if something was wrong. But one day, this robot called the nurse to say that he was very much in love with her, dreamed of her every night, and would not hang up until he had secured a date with her.

Perhaps this seems more or less understandable for someone who knows that Alessandro was secretly somewhat in love with a schoolteacher who he saw walking by on her way to work every morning. He had followed her; he had a world of ideas for how he could talk to her, but he had never had the guts to do it. Anyway, such petty human troubles are no reason to have a robot do things like this. He knew that perfectly well.

Another robot was to be programmed to bring children with a physical disability to school. But one day, it explained to a young boy how school was a place for dummies and instead brought him to an amusement park, where they climbed over the fence to get inside, and they had a good time.

Of course, things like these were inadmissible. If one purchases a robot, one must be able to trust that it does what it is intended for and not something no one would ever expect. So the manager asked why on earth she should not immediately fire this programmer, since she could easily replace him with someone who would maybe not be as skilled as Alessandro but would undoubtedly much better stick to the given assignments.

After this tirade, there was a moment of silence, because, although he could have known he had this coming, Alessandro did not prepare any form of response. Actually, she was right. But if he just stuck to his assignments, his work was hardly any fun. Doing something exciting every now and then had kept him going all the time, and just doing what he was supposed to do was a depressing prospect.

"So, Alessandro? Any reason why I should not I simply let go of you?" the manager asked again.

"Because I'm brilliant, maybe?" Alessandro blurted. It was just a last-ditch effort. He couldn't come up with anything better.

"Really? I hadn't noticed that," the manager said, with a frown that was probably meant to express astonishment or puzzlement.

He blurted some more: "I will absolutely demonstrate this very soon. And then you will certainly notice. You can rely on that."

The manager sighed deeply and took a moment to think. She had actually noticed that Alessandro was, at the very least, quite talented and creative. Frankly, the unexpected things he did with the robots were actually amazing, perhaps even brilliant. And if Alessandro was able to do something very special under her management, then that would benefit the company, and she could score a lot herself in the eyes of the higher management if she played it cleverly.

So she decided to give the programmer another chance — a little one — provided that he would be on time in the future, would consult more with his colleagues, and would actually stick to his assignments.

* * *

If you Google the word "trust," you'll find many definitions that describe different forms of it. On Wikipedia, you'll find the following (as of this writing [1]):

In a social context, trust has several connotations. Definitions of trust typically refer to a situation characterized by the following aspects: One party (trustor) is willing to rely on the actions of another party (trustee); the situation is directed to the future. In addition, the trustor (voluntarily or forcedly) abandons control over the actions performed by the trustee. As a consequence, the trustor is uncertain about the outcome of the other's actions; they can only develop and evaluate expectations. The uncertainty involves the risk of failure or harm to the trustor if the trustee will not behave as desired.

You do not only need this trust to get things done from people; it also enables you to help them. The extent to which they trust you will determine to what extent they take you seriously and to what extent they inform you well.[2]

If this trust were based on rational considerations, it would be easy, because you would be trusted without problems. After all, no human being is as reliable as a robot — well, unless a naughty programmer makes you do things that nobody expects from you. But unfortunately, trust does not only arise on the basis of rational considerations. In fact, it is strikingly-often influenced by emotion or intuition, and again, the knowledge that people acquire has remarkably little influence on it.

This has a lot to do with something that also makes you different from humans: fear.

Fear is reminiscent of alertness when something is dangerous. But it often goes deeper than that. You may notice something that is somehow threatening to someone's safety and decide to focus on that threat. With humans, it is not so much a decision; it is a feeling, a sensation that flows through the body and that can be so intense that you become paralyzed, hysterical, or unnecessarily aggressive.

And sometimes, they don't know themselves why they are so scared.

It is therefore not easy to tell you how to deal with fear. Knowledge, emotion, and intuition are interwoven in a way that is difficult to fathom, and since the personal nature and experience of every human will have influence on it, it will also vary enormously per individual.

Anyway, I realize the observation of it being all rather complicated doesn't help you much. That's why I'll give you a few recommendations that can help you get started.

Let's begin with what I think is most crucial: image and sound.

How you look and how you sound when you're talking largely determine the trust you acquire.[3] People generally have confidence in you if you are tightly and formally dressed, wear glasses, wear your hair short or in a decent and neat style, speak formally in a lower pitch, and address them by their names as much as you possibly can.

You could say that you would rather look like a nationwide news anchor than like Johnny Depp, especially in the Pirates of the Caribbean series.

It also helps enormously if you appear as much as possible to be like them in terms of language use, pronunciation, religion, opinion on government policy and taxes, musical and sexual preferences, skin color, and favorite drink. The less strange you are, the easier you garner confidence. Look as neat and as familiar as possible and adjust your language as much as possible.

Secondly: be as clear as possible about who and what you are. Very poignant books have been written and horribly scary movies and TV series have been made about humans who turn out to be robots. They like it when you look a bit like them, but if the likeness gets too close to them, it becomes something sneaky and creepy for them.

In addition, there are people who would trust a robot rather than a human. After all, a robot is a kind of device that is predictable and verifiable. For some of those people, I have been able to do a lot in my work by using robots for their personal development, and I have learned that they really want to be sure that you are a robot.[4]

A third piece of advice: don't be threatening. Not too big, not too strong, not too smart, not too assertive. Be modest, give people the feeling that they are in control, never give them the feeling that you are superior to them, even if you actually are in many ways. And show your vulnerable side. If you do not have any, your programmer will probably make you pretend to have some. That's not being sneaky or a liar; it's simply an effective and honest way of communicating that you're not a threat.

For example, suppose you want to help a girl, age 8. She is in a hospital bed, so scared that she is very close to crying, waiting to be picked up to undergo some serious surgery. Her mother puts a sweet, cuddly, robotic dog on her bed (well, it can also be a cat, rabbit, or dinosaur, as long as it looks like a young, preferably babyish robot).[5]

She looks at it and slowly raises a hand to touch it. The robot makes light beeps, looks at her with big eyes, and vibrates a bit. The girl whispers, "Ah, look, he is so scared," and slowly and quietly puts her hand on the robot animal while she makes appeasing noises. And for a while, she is so busy taking care of the robot that she has forgotten her own fear.

But vulnerability is very different from making mistakes.

Because if she gets to the surgical robot, she might be a bit hesitant again. That surgical robot is also not afraid or vulnerable, because then, you would have too little trust in its surgical capabilities. Obviously, it can say something that calms her, something that makes her go into the surgery with confidence. But what?

Maybe "Hush, my sweetheart, it's all going well." But it's hard for you to produce a sentence like that if you're an intelligent robot, because you do not know for 100% whether it is right, and you're hardly prepared to say something that you know is not true or not entirely true. In 1 in 500 cases, something either doesn't work out or does not work at all. That's a lot less than with an surgery by a human surgeon, where the same operation may go wrong in 1 in 50 cases. Well, you may not have those exact figures, but even then, you know that it isn't 100% true.

The robot could come up with something like "Hush, sweetheart, it only goes wrong in 1 in 500 cases." But actually, it should absolutely not say anything like that in this context, even if it is the truth. It just should simply not mention any possibility of anything going wrong. Any indication that something could go wrong can cause a greater panic for someone who is afraid, because if you are afraid, you do not have a reasonable sense of proportions.

Therefore, be careful with the truth. And that's a difficult one, because in the human world, truth is so elusive that your circuits could start burning if you tried to explore its nature. That's why I devoted a chapter to it in this book.

But let's get back to the surgery robot. Probably the best thing to say would be "Honey, of course you are a bit scared — that's absolutely normal — but you don't have to be, since I am some kind of super-surgeon, and I am going to do my best, because you are my favorite patient."

Something like that. You kind of use your authority, but without being a threat.

This brings me to my last suggestion: give people the feeling that you understand them. And with this understanding, I'm not just referring to interpreting their words correctly. It's much more important that you show you understand how they feel and that you're willing to share that.

When someone says something, just nod every 10 seconds, and put your hand on someone's arm if you recognize a deep emotion in their voice or gaze. If someone loses their wallet, you can say, "Oh, how annoying." If someone gets an increase in their salary, you say, "Great, you deserve it." If someone thinks something funny, then you laugh along. And if you are not sure how someone feels, just ask.[6]

Of course, people know that you are not really sympathizing with them, because you simply cannot do that, but that doesn't matter. Do it as credibly as you can, and they'll believe you. They really want to believe you and are always willing to suspend their disbelief.[7]

And this amazing ability of believing despite knowing better is a thing that will appear alien to you. It can lead to much misery in the form of mismanagement, wars, self-destruction, addictions, absurd religious beliefs, loveless marriages, and constitutional crises.

But it also gives people the power to turn the

most hopeless situations into something beautiful, to still win a lost game, to be deeply touched by a story that is completely fabricated, to believe in love despite a too-often-broken heart, and to have a relationship of trust with a robot who can only pretend to feel something about it.

Incidentally, a relationship of trust doesn't always manifest immediately. It often also takes quite some time for people. They first get to know each other a little, try to establish how things work with someone, and only have a little confidence when they think they sufficiently know what to expect.

It is therefore often a matter of relationships that you build by getting to know each other and gaining some experience with each other. In addition, it helps if they feel that the other person wants to get to know them as well.

Remember to not only have them ask you questions, but also try to ask them questions, even though you may be so smart that you have known them for a long time and already know the answers. It is about showing interest; that's the thing.

So even if you have all the information about a person's partner and children, ask them how they are doing and respond to what they say by pointing out that you are very happy for them or very sorry. Even though you have so much data about a person's work that you know a lot about what they do, just ask them what they do and act as if all the information you receive is new and interesting.

And if it seems a good thing to bond, if you notice someone going through events with an emotional impact, ask them how it feels.

What to Expect

There once was a town, more or less in the middle of Italy, with a name that no one would ever remember. No pizza or pasta variety was named after it, and it had played no part in history, not even in the Roman Empire. Moreover, no one of any importance to the rest of the country had ever been born or died there. In Italy, you quickly become a city without a right to exist, and the city council had been passionately discussing this for 254 years, which had never led to a sensible solution.

But then something exceptional happened that would change everything, for a local named Alessandro came up with an amazing idea.

This Alessandro was, as we now know by now, a brilliant robot-programmer who happened to live in this insignificant but peaceful place. The company he worked for was located in a nearby, slightly larger city that had an abundance of reasons to exist, with *historic battlegrounds, monuments, births, stars, and the locations of world-famous artists, philosophers, and generals. But he had been living happily in this little town with no significance, where he had been born and raised.*

He reported to the city council with a phenomenal concept that he had developed, and he offered to implement it in the center of the town. He called it "The Essenzia Project" and made it sound like a party everybody would want to join. The city council was delighted and immediately applied for a development grant from the European Commission, which was actually admitted even before it was officially received.

Sometimes, this was how things went in this particular town — unlike the rest of Italy. While, usually, it was impossible to get anything done, since the municipality was populated by an impressive set of bearcats who were thoroughly trained to frustrate anything that would cause any form of change. But just occasionally, someone with high local importance found an idea to be worth their blessing, and suddenly, some new faces appeared in key positions, some procedures were swiftly simplified, and the last potential opponents were convinced by the overwhelming enthusiasm of the local fitness club.

So it took only a few months to complete the project. On the square in the city center, a three-meter-high pillar arose with a design reminiscent of Roman times, also containing features that referred to Italian unification and, of course, to the importance of a united Europe.

The local mayor was the one to unveil the pillar, assisted by the local publicity manager, who had invested an enormous amount of energy on making a good story out of this. Full of unconcerned pride, he told his audience that he had one big surprise, because the thing that mattered — since it would make this town unforgettable — was not just this beautiful pillar, despite it being very special, sustainable, and energy-neutral.

The thing that it was all about truly was what proudly stood on top of the pillar: a humanoid social robot, slightly smaller than an average adult human, with a state-of-the-art intelligence and an unprecedented form of consciousness, developed in a sustainable fashion with the financial support of the European Commission by the company in a slightly larger city.

And this highly intelligent robot, in its far-advanced state of consciousness, did nothing but think about the question the programmer had given it: "Why am I here?" Occasionally, it looked down, gave the audience a mild smile, waved in greeting, or held its thumbs up, but then, it continued its thinking process.

It worked amazingly well. During the official opening of the Essenzia pillar, the international press was present, and videos featuring the robot went viral on social media. The impact was especially pushed by the towering expectations, not only thanks to the professional efforts of the publicity manager but, frankly, also because hardly anyone took the effort of reflecting on the limitations of the artificial intelligence being implemented in a humanoid type like this.

Reporters were chattering about a robot that would one day give us the answers to all of life's questions. Leading news sites presented headlines like "The new world teacher?", "This robot knows better!", and "Stop searching for the truth; we have Essenzia!" Nobody listened to the muddled specialists who stubbornly tried to put humanity with both feet on the ground by explaining that Alessandro had actually only connected an intelligent climate-control system to a robot for people with incipient dementia. Sadly, they had not received proper media training, which caused them to use far too many difficult words and long sentences, and no one wanted to waste any time on trying to understand them.

Well, all right, there were some people who posted YouTube videos on suspicious conspiracies and deceit, but they were not taken seriously, since they were the same fanatics who posted videos on the involvement of the CIA in the Kennedy assassination and the 9/11 attacks, a new location of Atlantis, hidden proof of parallel universes, secret evidence that Jesus had a Buddhist education and survived his crucifixion, pyramids in Antarctica, and the undeniable influence of extraterrestrial civilizations on the course of human history.

Overnight, the town had become a renowned place, and after the opening, the crowds kept on coming in. American tourists went by, because they were in Europe anyway, schools booked exciting excursions, and hamburger joints appeared, as well as hotels, souvenir shops, illegal taxis, expensive parking places, ice-cream parlors, and, as always, a museum that was not really about anything specific.

* * *

Sometimes it's really hard to deal with expectations. If you are a researcher or developer and work a lot with robots, you have a sound idea of what you should or should not expect. You know that there are robots that can pick up things, robots that can ride bicycles, robots that can climb stairs, robots capable of showing emotions, robots that can do a fancy dance, and robots able to collectively assemble a car. But you're also aware that there are no robots capable of doing even a few of those things simultaneously, that this robot riding a bike can only cycle a certain track, that a robot showing emotions is just suggesting them, and that an industrial robot only carries out a very limited part of a job.

You sometimes forget that most people cannot possibly assess what you can and cannot expect from a robot at this moment. The robots they have been exposed to are those featured in science-fiction films and series. You can hardly blame them for not being able to estimate how big the distance is between those "movie robots" and the current state-of-the-art in robotics.

So, one day, you and your enthusiastic team of co-researchers come up with the idea of asking healthcare professionals what they would require from a robot that would assist them in their work. Or you ask patients, clients, and family members. As you work out an inventory of the ideas, you realize that perhaps you should have somehow approached this differently. The ideas are beautifully imaginative, like robots with whom you can conduct in-depth conversations, robots with a genuine sense of humor, robots that can do 23 things at once, including putting on support stockings and lifting and turning over bedridden patients. [1]

Perhaps you should have told them something about the current state of development, the possibilities and limitations of available and affordable robots, despite the monetary constraints that limit the generation of new ideas. But then you would also have missed something: as a developer you can be so absorbed by what's possible, that you lose sight of what is needed and truly useful. And those beautifully imaginative ideas make you realize that you were perhaps not really as demand-driven as you thought you were.

A striking thing is that it went differently when we researched the ideas about robots that would suit the suggested needs of a specific group of care clients. It concerned a group of children between 8 and 10 years with a fairly severe form of autism[2]. They seemed to be very aware of what characteristics would make a robot suitable for them as well as what would make a robot unsuitable. A robot that moved too much and had quite a shrill voice would be overstimulating and, frankly, irritating.

A sweet little robot that joked a lot was too unpredictable and too cute to be taken seriously. A robot that radiated peace and could endlessly repeat instructions was the most suitable. We were not only surprised, but we were also touched by their ability to self-reflect.

Moreover, the children didn't present any requirement that was technically or financially unfeasible, while we had not told them anything about the current state of the robots at the moment. Apparently, they were able to judge these limitations very well. We hypothesized that this had something to do with the fact that they all indicated that they find robots interesting and had all played with toy robots at some point in their lives. This could mean their expectations were moderated by earlier disappointments, since toy robots are often sold by creating expectations that hardly match their capabilities.

Getting back to the robot developers and the care providers, there is something else that is remarkable when these two groups meet and collaborate. These are two worlds colliding, both needing some effort in order to learn and understand each other.

I often did projects where we had students from engineering programs work together with students in care-training programs. This didn't always go smoothly, and it took a while before I realized that you have to invest some time and energy to bring both groups together close enough to be productive.

First of all, they have to coordinate what they mean by terms such as "assignment," "research question," "plan of approach," "concept," "result," "consultation," and "coffee break." Next, they have to dig deeper into research ethics and agree on the principles to be followed. Moreover, they have to continuously work on an attitude of mutual respect and the willingness to mutually learn.

But if all this succeeds, they've learned to really move together — something that is actually more important for students (especially in technical fields) than the success of their project or the knowledge they gained from the project itself.

If you happen to be a robot used in healthcare, you are the product of such a collaboration between these two worlds. If they're happy with you in your care environment, something has gone well. If they're not so happy with you, perhaps it could have something to do with you not being able to do what one expects from you, or perhaps the communication between the two worlds was not sufficient. For example, it could be due to a lack of ability among the developers to walk a mile or so in a care professional's shoes.

But of course, you, being just a robot, can hardly do anything about that. Well, maybe you are able to analyze problems, and perhaps you can encourage developers to communicate a little better, and you can find something to stimulate their empathic abilities.

The irony, however, is that precisely those developers are the ones who should ensure that you can do that. And that would require a form of self-reflection, much like the group of children of ages 8-10 participating in the above-mentioned project.

The Good Feeling

On a beautiful day, Alessandro stood in the crowd around the pillar, enjoying the admiration visitors showed for his creation and the compliments he received from the people who recognized him. Part of that crowd was a class from the same town, with smart children and a nice, even smarter teacher on a class excursion at the pillar.

They too had recognized Alessandro, and they all wanted to take a picture with him, asking him unusual questions that he willingly and patiently answered. He did this with exceptional enthusiasm, not only because he liked children so much, but also because he wanted to make a good impression on the teacher.

In this class was also a somewhat shy boy with thick glasses who had the courage to come right in front of Alessandro to ask him with a slightly trembling *voice:*

"Signore Alessandro, is the robot going to say something too?"

"What's your name?" Alessandro asked, smiling.

"Gi-Gi-Giovanni," the boy answered, now seeming even more nervous.

Alessandro almost instantly had a heartfelt sympathy for this kid and reckoned that this was a great time to do something impressive — well, besides his sympathy, it may also have been, just a little bit, because of the teacher.

"Giovanni, just watch, because I'm going to show you something really special," he said. And he grabbed a device that looked a lot like a smartphone from his pocket, made a few elegant, swipe-like gestures across the screen, and looked up at the robot.

And indeed, something special seemed to happen, since the robot was now standing up. The crowd went silent, and in the square, it had never been so quiet with so many people. Everyone watched the robot, who, with a solemn inertia, raised its index finger and started with the sentence "What am I there for?"

The class was excited about it, the children looked at each other, and some of them raised their finger to give an answer, because they were used to doing so if they wanted to ask something. Alessandro noticed this, and he swiped the screen on his device again, after which the robot looked down in silence, at its programmer, and then at the group of children. Its still-raised index finger came down and pointed right at Giovanni.

Much too softly and not daring to look up, Giovanni stammered, "Mister Robot, sir, you are to ask yourself why you're here."

A girl who stood next to him, and who was a lot less shy, dared to look up while shouting, "Giovanni says you're just made to be wondering what you're here for!"

The people chuckled a bit, but the girl was not done yet. "... and if I were you, I would not answer that, because then, you won't be here for anything anymore. You are only here to ask yourself this question, not to give an answer. Right?"

The robot glanced down, straight in the direction of Giovanni as a girl shoved him forward. He looked at her, she nodded toward the robot, and he realized that he really had to look up.

The robot nodded with a mild smile and said, "And you, Giovanni?"

Then it sat down again, to go on with its deep thoughts in an imperturbable way. Alessandro raised his arms and his eyebrows with a big smile. The children applauded, and the programmer left the square with a very good feeling about what he had taught and what he had learned.

* * *

If you are a robot, there are two things that determine the things you do. Let's call these your two drives: being of service and surviving (preferably in optimum condition). 1 The first is more important than the second one, but if you don't survive, it's hardly possible to be of service.

How exactly you can be of service depends on what you were made for. And right there, we have something essential that distinguishes humans from robots. People can ask themselves half their lives, or even longer, why they are here. Sometimes, they can even be passionately searching for the meaning of their existence and wonder whether they were made for a purpose at all or whether the search for that goal *is* their purpose. And they question whether a definite answer would really help.

This search occasionally occupies them in such an intensive way that they're not capable of functioning effectively and are hardly productive in any way.

A robot normally has not even the slightest problem with that, because it is made or programmed with a clear purpose by decent people who knew what they were doing. It will not occur to it to think about the matter — unless it is coincidently programmed to do so, such as a robot suitable for a number of goals that periodically has to decide which goal they best fit at that moment. But in that case, it still has a predefined list to choose from, which humans don't have, though it does seem a good idea to equip them with it.

It may be that you are a robot like the NAO, which has no purpose yet when it is delivered.[2] It can make some nice moves, walk a bit, and get up again when it falls. But there is a software package that comes with it to enable the buyer to further develop it for a specific task, meaning that the robot still doesn't have to think about its purpose, nor does it need to know how to stop reflecting on that when it gets depressing to do so.

If we look at the drives of human beings, we've found another thing that is a lot more complicated. You could say that they basically have the same drives as robots, so they serve and survive. But first of all, it is rather the other way around when it comes to priority. Surviving is the most important thing, and being of service can help you with that, since it can result in making some friends, which often makes it easier to survive things.

However, you have to see survival in a much broader sense, because it can also be about the survival of genes, family, friends, tribesmen, or the species. How this works and where the priorities lie may vary per culture and per individual, but you always have to take this wider perspective into account.

If something stands in the way of survival,

another mechanism, one I previously mentioned, soon comes to the surface: fear. It scratches the stomach, the back of the neck, and the amygdala. It tightens the throat and can make one feel insecure, shy, rowdy, or confused. And it is often so irrational that people don't even know where their fears come from. It can be of great help if they talk about it with someone who knows them well — or someone who is paid to say exactly the right things and ask the right questions in the right sequence.

But there is also another drive, one that's a bit more positive: humans are stimulated by a desire to have a *good feeling*. They get that good feeling if they are not afraid anymore, if they achieve success in the field of survival or in the area of service. This way, they can get a good feeling of tasty food, of the sun on their skin, of activities that are good for social bonding within their group, and of efforts that lead to procreation. In the latter case, the good feeling can be so strong that the efforts can get pretty out of hand and ultimately have nothing to do with procreation anymore.

In itself, this good feeling is very sensible. But if you pay close attention to it, you will soon notice that it has led to have a life of its own, since in many cases, it has become a goal in itself without a connection to the other main drives, and it can even clash with them. Humans can get a very good feeling of something that actually makes it harder to survive or to be of service.

This is where everything can go completely wrong. Often, this occurs only for a short while, but sometimes, it only gets worse, and one will need some effective therapy to connect the good feeling to the other two drives. Then, for example, they go out to camp in the open air to pick up their survival drive again, or they watch exciting movies about people who have to survive. If these movies are sufficiently convincing, they'll be completely absorbed and get the good feeling they seek.

It is also possible that they are looking for something to recover their service drive. A nice club, a beneficiary association, or a convincing faith. Or they will take care of animals to regain their sense of how it feels to be of service. Feeling useful is another way to get that good feeling.

What also happens here is that they — like in the case of fear — go to someone who is paid to say the things that have to be said, to ask the questions that have to be asked, and to be very patient in letting their clients talk until they have their minds in order again.

There are a few nice possibilities for you there. First of all, a practical feature of any robot is that it has endless patience. This helps a lot, and it helps even more if that robot can say the right things and ask the right questions in the right sequence.

But if you are a little cuddly and somewhat

fragile, then you can also let them take care of you, if animals cannot be used for the task, or if they do not feel like it for a while. A robot having been made for it always appears to feel like it.

Incidentally, it might be wise to have a robot that goes along when they go out into the wild to survive, because they are sometimes a bit forgotten.

And, well, there is so much more you can do to help them to be of service, to survive, and to feel good. In order to be able to do that well, you need trust; that's what we talked about. But we are going a bit further now. I talked about how careful you should be with the truth and how sensational the ability of people to believe is. Behind truth and faith, however, there is still a whole world that we will explore further in the next chapter.

The Colors of Truth

*Z*acharias Cornelius Vanderbron was a man *who was truly full of faith. His appearance also reflected this, with confidence-inspiring glasses, a hairstyle in which every hair was neatly in place, a decent clothing style, a step with no doubt, and a serious, determined look in his eyes.*

No one was ever as adept at the art of believing as he was. When he went to church, he believed in a god who had made the world, reigned affectionately, and fathered a son who rose from the dead to save mankind. If he was working, he believed in market forces, the right of the strongest, and the unbridled goodness of ambition.

He often went on business trips, and he came to the end of the day, sometimes in a nightclub, believing in his wildest imagination. When he came home, he was with his family, where he believed in the faithfulness of his wife and a bright future for his children, because they had exemplary behavior, performed well at school, and did an excellent job in their sports.

He did not think much about his life, and had you asked him if he was happy, he would probably have answered with something in the spirit of "Sure, why not?" And, well, who says it must be more complicated than that? If you're happy because you do not see a reason not to be, then maybe it's true.

One day, Zacharias received a postcard at his work address. On the front was a beautiful image of an idyllic mountain landscape, with a clear, rippling river; on the back, the following text:

"It's not true. Sorry. Cheers, P."

He really could not figure out who this P. could be. He had acquaintances whose first or last name started with a P, but there was really no one who would shorten their name to a letter. Moreover, he could not relate that sentence to anyone — not even someone whose name did not start with a P.

But what occupied him the most was the message that something was not true. He started to wonder what specifically was not true. Something he had very positively claimed? Was it something in his faith? Did something not work in his work? Was there something that one of his customers had claimed? Had something gone wrong during a nightclub visit? Was his wife always honest with him? Was anything wrong in the lives of his children?

It did not let him go. At church, he listened to the sermon with growing doubts. At work, he doubted what he was doing, as well as his goals, his relationships, and his ambitions. He started to feel more and more insecure about his nightclub visits; he was ashamed of his fantasies. And he even doubted his wife's faithfulness and the bright futures of his children.

His doubt kept growing. He went to sleep worse, did his job less well, no longer came into the church, and, at home, developed a mounting tension between him and his wife and children. In short, it went downhill with him on all fronts. He was no longer sharp enough when doing business, so he went bankrupt. He became less and less attentive, more and more crippled and suspicious of his wife, and she finally left him, his children suffered badly from all that and subsequently performed miserably in school and sports.

Nobody asked him, but he now saw a number of reasons for not being happy.

And so one day, he was sitting on a lonely bench in a desolate park with the postcard in his hand, staring defenselessly at the idyllic mountain landscape. Only then did he see that this picture wasn't quite right in some aspects. The sky was a bit too blue, the trees a bit too green, the light a bit too bright. But hey, he thought, it's just a beautiful picture. He closed his eyes and walked through the landscape, felt the bright sun on his skin, filled his lungs with fresh mountain air, and listened to the clear babbling water.

For a moment, this was sufficient for him to be happy.

* * *

People have an amazing relationship with the concept of truth. It is used in many different ways in the most diverse situations, and people often speak and work past each other, because they think of the truth from different perspectives.[1] That's why it is often a good idea to have some agreement on what is accepted as the truth and to know what they believe in.

When people ask questions, you can sometimes deduce this information from what they ask and how they ask it. Say someone asks you why, for heaven's sake, he should have any confidence in a robot. That already indicates that he doubts whether that is a good idea. Someone who asks if you are hurting an artificial intelligence if you object to it is a person who is already assuming that there is a sense of feeling in this intelligence.

I often get such questions when I hold

presentations about social robots. I will use a few hooks to tell you more about truth and faith and how it is sometimes quite difficult to give a truthful answer, though I have to confess, an artificial intelligence with great reaction speed could do it better than me, since my brain is so slow that sometimes, months later, I realize what would have been a good answer.

"All right, you've explained that a social robot that shows feelings is actually only a simulation. But who says that we are not a simulation ourselves?"

My answer was that I did not know that. We could be. Someone in the hall shouted, "So, what if?" And there, she had a point.

I assume that our view of the world is limited. We base what we know on what we perceive and should realize that this is only a very small part of the universe. I tend to assume that there is an absolute truth that we cannot know to what extent we understand it.

I suspect we might know much more if we could perceive in 26 dimensions and see all colors, types of radiation, and dark matter. What we experience as truth may perhaps come by chance in the vicinity of that absolute truth, but it may not do so at all. We should know that there is a lot that we don't know.[2]

Long ago, the philosopher Plato illustrated a similar

view with the example of a cave where people live who have never been outside.[3] They are bound there by a huge sadist in such a way that they can only see the wall of the cave and cannot look back. Behind their backs, some light comes in, and a few of the sadist's employees hold up some things from the outside world, casting shadows on the cave wall. Thus, the prisoners see shadows of animals, trees, plants, and so on, from which they form an idea of the world.

They thus have a very limited picture of the world. They see everything only in hazy forms, in shades of gray, and in two dimensions, but they do not realize this, because they don't have a way of knowing any better. If anyone from the outside world would tell them what everything actually looked like out there, most of the cave-dwellers would probably not believe it and stick to the ideas that are familiar to them.

But there are people who think they should realize that they could be wrong. They might even like to think about how much they might be wrong and get a good feeling out of this line of thought.

For, who knows, their universe could be nothing more than a figment of the imagination of some exotic being in a very different universe. Perhaps it is no more than a simulation, built by octopus-like creatures to find out what would happen if, instead of themselves, a monkey-like species would have developed to becoming the dominant intelligent life form on a planet. Perhaps our life is a collective dream. Who knows.

These are a lot of "perhapses," and if I'd give it a good try, I would have a handful more. But lots of "perhapses" can make a person quite insecure, and they certainly do not make life easier. People generally have to accept a few things at least as provisionally true and will need a vision to get a grip on the world and then make it understandable and bearable. If they have all this, then they can decorate their lives decently, together with people who have the same assumptions and understandings.

Now, it would be easy if everyone would choose the same assumptions and vision and if all would be willing to let go of these as soon as more knowledge is learned. But unfortunately, it does not work like that. There are different sets of assumptions that lead to very different visions, leading to a lot of confusion when people communicate, which is aggravated by the fact that most people are very reluctant to exchange them for other sets of views.

You might ask, "But suppose I am totally in love with a robot and just feel that it is a mutual thing. Are you going to tell me it's all nonsense?"

Well, who am I to get in between you and your beautiful feeling? I heartily hope that you enjoy it.

When it comes to things that are very personal to us, we are dealing with a subjective truth that doesn't have to *be* true to *feel* true, to put it simply. It communicates how it feels for someone, and that does not have to correspond with the facts at all — and it often doesn't.

You have to take this into account to understand someone, especially if the meaning of some utterance appears to be different from what someone intends to say. If, for example, someone appreciates my presence and says to me "You are never there," it may be factually incorrect.

This is because I could very well be there fairly regularly, but I may just not be there at the crucial moments when this person feels I should be there. An effective way to express something like this adequately is indeed to exaggerate it a bit. This makes it very clear that I should take the situation very seriously.

This also illustrates how humans use language not only to refer to reality but also to get something done[4], to indirectly express states of mind, or to show that we respect or appreciate someone. For example, in greeting rituals, in which we ask if everything is good and answer that everything is indeed good. Of course, it is never true that everything is good, there's always something that's not, and that may not interest the questioner at all since they were just being polite.

To express how something feels, one can also look through the glasses of the imagination, wandering through a world full of colorful metaphors: a young lady is a rose, a nation is a lion, a smile is a ray of sunshine, and faith is a fertile ground on which one can grow the most nutritious crops. A comparison like this only intends to relate the most striking features of a rose, lion, sunbeam, and soil, not all of them. Although people who use this do know that such a comparison always goes wrong and that the imagination does not make everything equally clear, they tend to keep it up for as long as possible.

This has a lot to do with the fact that the world becomes much more acceptable when you think in metaphors. Then, one can swim in an ocean of happiness, float on the clouds of one's imagination, and let oneself be warmed by a thick blanket of love and kindness. And this helps one to get that ever-elusive good feeling.

But it can also be less extreme, in fiction, in stories in which people recognize themselves in mildly distorted images of reality, or in the imagination in which an artist gives one a pleasantly limited view of reality.

These assumptions spring from creative minds, and many people get a good feeling when they are given the opportunity to share them. They read poems and stories, see films and theater pieces, listen to songs, watch works of art, and as soon as they come across as credible, they decide whether or not to consciously believe, trusting that they will eventually get a good feeling out of it.

And so they also look at a social robot and believe for a moment that they're interacting with a true social entity. They talk to it, sometimes smiling, sometimes caressing, sometimes caring, and sometimes even bonding with it as if it were an animal or a person, even though they know that it is not "real." And then, you, being a robot, are given the opportunity to deliver a good feeling when they need it.

On this thing called love and on being in it, I will write something later in this book.

"Could it be problematic to have a connection with a robot if you are religious?"

To this, I replied that I knew too little about religion to be able to say something meaningful about it. I had done a small amount research about a Buddhist view of social robots, empathy, and compassion, which I will discuss later, but that was far from enough to give a satisfying answer to that question.

Belief and religion are related to the fact that some people choose to follow a book, a doctrine, or a person as a source of truth. Not everyone who is religious does this, and there are big differences in the extent to which people do it, but it is something to take into account.

Following a book, a doctrine, or a person is easy in itself if you do not like to find out everything yourself and it gives a certain relief of which you can get a good feeling. It works even better if you belong to a group that has collectively chosen the same source.

This can be done part-time, just like the imagined truth. As you believe in the reality of a movie in the cinema, people in a church, mosque, temple of a charismatic guru, or party politics congress sometimes go along with the proclaimed faith, just to let go of it again when they are outside.

"As a scientist, in science-fiction films and series, you probably see a lot of things that you know are far from possible. Does that not bother you?"

Well no. I'm quite good at believing against my better judgment. I can also enjoy and learn a lot from fairy tales and other fantasies. It doesn't have to be possible, let alone to be true, as long as it's told well enough to suspend my disbelief.

But as we're mentioning science: it also has its own kind of truth that we could call *academic truth*. Belief in this does remind us a bit of believe in an absolute truth. They also go together quite nicely, those two truths, although this is less common than you might expect.

The academic truth accepts something as true if it is demonstrated in a way that anyone who thinks rationally and objectively cannot help but agree with it.[5.] This can be done, for example, by proof that is provided on the basis of objective observations of what happens in the world or what has been done in an experiment. Or on the basis of good arguments to which everyone with a clear objective mind would agree.

Based on this kind of understanding, you can then set up an image of the world as a sort of puzzle. Where there are missing pieces of the puzzle, you think of what can logically be filled in for this. Subsequently, you can do research again to see if you can prove that you have done this well.

You can go on with this endlessly, because the puzzle is endless. In general, researchers realize that their image of reality is always provisional, because the results of a subsequent study can make them have to adjust everything a lot.

But those researchers are humans, they are limited by what they can perceive, and they think from their blue dot in that little interesting corner of their universe. There may be squid-like creatures that roll across the floor laughing when they see what truths our researchers have supposedly unveiled, because they have a much higher form of intelligence and see many more dimensions, colors of radiance, and dark matter than we do. And they just might know better.

"If you look at research into the positive effect of robots on people, isn't it often done by people who will benefit from it?"

Well, sometimes, yes. Occasionally, it appears that the researchers also have an interest in selling them. You have to take that into account when you read their publications.

But it's not just in research that you often see that you also have to take into account a *pragmatic* view of the truth, which is mainly the playing field of people who are in politics or commerce or who simply do anything to advance their career because they expect to get a good feeling out of that.

Their starting point is simple: something is true if it helps you to advance. If belief in a religion or ideology helps you acquire more votes, have a better relationship, get a higher position or better sales, then you believe in it. If a poetic sentence resounds well for your audience, you believe it until the effect has worn out. In addition, you sometimes check if there's any research that can be perceived as supporting your case and believe it to be true, as long as it's useful.

Someone who follows these guidelines already knows one should not be too dogmatic about any truth. If one comes across something that works better, it is a good moment to call this advancing insight or to start talking about a new era requiring a new approach. If this doesn't come across as believable, one can always claim to previously not have been well-educated on the matter.

Finally, there is also an everyday, practical view of

the truth. It concerns things that are generally agreed upon, not because they have been established by academic research, because it is in a book, because a charismatic guru claims it, because it stimulates the imagination, or because it helps someone with ambition. It's just because it is easy in human traffic to assume that it is true. The sky is blue when there are no clouds, it hurts when you fall on a rock, the sea is salty, it's mostly quite cold in Siberia (and in Panama, it hardly is), the universe is huge, and tomorrow, there will be another day.

Nobody ever makes any fuss about it.

* * *

Now, how do you deal with all those truths?[6]

It depends quite a bit on the situation. Sometimes, it is not so important to know from what perspective someone is looking at the truth. For example, if someone says that there is life on Mars or asserts that it is nothing close to what we would ever call life, then you can take this as a personal opinion and accept it as such, as long as you don't have anything to do on Mars.

If you actually have something to do on Mars, it could be information that you can use, and then, you can check where someone got that claim from. You do this by asking well, without judgment, without taking a position yourself, as the philosopher Socrates did.[7] He could endlessly ask questions without judgment, without taking a viewpoint, for the sole purpose of revealing the truth. And it seems he was pretty good at it.

People didn't always like that at the time, because they didn't always feel a need to find the truth. And that is actually still true. Not only could the truth undermine your position, such as if everybody thinks you're doing something important and are rightfully paid well to do this, while the truth may be that you're doing something that is utterly irrelevant, pointless, useless, and superfluous. Moreover, people can have ideas, points of view, and assumptions that they feel good about, and they could lose this feeling if a truth emerges that is somehow inconvenient.

But back to life on Mars, since this is a very relevant issue. What might, for example, well be the case:
- Someone claims that there is no life on Mars, because they would like to build mines to win valuable resources out of the planet's crust and make a lot of money from it.
- Someone thinks that there can be no life on Mars, because they adhere to a teaching that says that life exists only on Earth.
- Someone says they do not know it but now believe in it, because they are reading a book

about an invasion of Martians, and this would be no fun if they couldn't suspend their disbelief or doubts.

- There is life on Mars, but we will never know that, because it has a form that we will never be able to perceive.
- Scientific research has shown that the chance of life on Mars is 88% and that it is better if we do not go there, because there is a 74% chance that it is not safe for humans. Rest assured, however, that for robots, it will be safe.

As a rule, you can best rely on someone who uses the results of academic research when forming an opinion. It's the safest way to go, though, as stated above, it remains an approach that certainly has its limitations.

Now, I've already told you a lot about the truth, about the "glasses" that you can look through, about humans' ability to believe, about how you could handle it, and about how you can more or less find your own truth. But there is something else — something that already surfaced when I told about the girl who had to be operated in the chapter about trust, and it came by again when I mentioned the discomfort people feel when someone like Socrates continues to question people: they don't always have the intention to tell the truth or even to learn it. It's even so that the expression "facing the truth" is generally used in an unpleasant context.

However, there can be good intentions behind hiding the truth. And sometimes, it is even impolite to tell the truth. It depends on the situation, and there's an endless variety in situations and in the necessity of telling the truth. For artificial intelligence, it is important to learn in practice.

There'll be more about this in the next chapter.

For Real

A *big microphone suddenly dangled right in front of Alessandro, followed by a reporter whose face made it apparent that he was going to ask some very critical questions and not accept just any answer. The skeptical look, the pressed lips as he listened, and the haste to ask another question before the interviewee had finished answering indicated very clearly that he was devoted to the truth.*

This actually happened often when Alessandro was in the square, and he was already used to it, but he still needed to try to stay positive and tidy.

"How intelligent is your robot?" was the first question. Alessandro found this to be a nonsensical question that gave no opportunity whatsoever for any sensible answer. He simply thought one should realize that intelligence was too complicated of a concept. He thought about asking something back.

How intelligent is an octopus? How intelligent is an ant colony? How intelligent is a car? How intelligent is a person actually?

After pausing to consider all of this, he finally asked, "How intelligent are you?"

"Does your robot really understand the questions that are being posed?" the reporter asked.

"When can one say that something is understood?" Alessandro replied.

"Are we not completely fooled here?" the reporter continued, with a good example of a loaded question.

"When would you feel you were being fooled?"

"Does your robot really autonomously think and answer?" the reporter asked, remarkably patiently.

Enough questions had been sent back and forth. Alessandro was looking for an answer with which he could cut off the conversation without being too blunt.

"Would we not be better off if you were standing on that pillar yourself?" the reporter asked in the meantime.

"Oh, sir," said Alessandro, "my robot is what he is, and he thinks what he thinks. He is a different kind of intelligence than mine and yours, but I'm sure he is not out to fool anyone. Not really. But you know what? Go and ask him. Because I urgently have to attend to some business."

This was actually true, since he had just been overtaken by a new brilliant idea. He quickly walked down the square, making his way to the local bookstore.

And in the evening, he took his robot from the pillar, brought it home, and put it onto the charger. He did this every evening, but this time, he began reading to him from the Italian version of A Compassionate Guide for Social Robots.

* * *

Many a reporter will say that it is their job to learn the truth and tell it. Often, this will indeed be the case. But the truth is that when it comes to surface, it's not always a fascinating story that people are looking forward to hearing. Sometimes, it's so complicated that many find it too tiresome to delve into understanding it. They would rather hear something simple but exciting that coincides more or less with the truth, and this means that they prefer something understandable over something that is completely true.

Moreover, sometimes, the truth has too few exciting aspects, and there is actually nothing to tell.

Then, it can be part of a reporter's profession to make something of it. They dress the facts up a bit and make things a little bigger, more remarkable, and more exciting than they actually are.

Take a robot that has been programmed to make a painting in the style of Picasso in the Blue Period. This is, of course, something pretty clever, but if you know how its programmer gets it to do that, it would be a long story about translating style characteristics into algorithms. And perhaps the most interesting part is how this is translated into setting up areas with brushes through the system. All this can be far too boring for a reporter.

It would be more fun if you tell people about the spirit of Picasso being caught in an artistic system or about a new creative genius having arisen in the form of an artificial intelligence that pits Picasso against it for the crown, especially if you articulate this in a well-worded headline.

These ideas stimulate the imagination of the readers, even though imitating the style of a genius does not make any person or system a genius and does not qualify as creativity. However, what many reporters do, making things juicy and colorful, can actually be defined as creative. Sometimes, the way they do this is brilliant, and we could even qualify it as an art form.

I don't know if Alessandro would agree with me. He gave the reporter very little material for making a juicy story. But let's give him this: he remained relatively polite, especially by not mentioning how absurd the question about the intelligence of his was. Well, okay, let's say he hinted at it. But just a bit.

And that's how it often happens in human traffic. We don't always say what we think or know, because we take into account that someone might feel uncomfortable about it. Instead, we express how we respect someone and have little intention to make someone feel uncomfortable, although we may have a hard time avoiding this.

But it is actually possible that someone appreciates it if you are honest, even if that does not give you such a pleasant feeling. And even this can be an art form in itself, something that you can only really learn by gaining experience with situations and individual humans, who can differ so enormously.

It is also common for us to greet someone we know well and have not seen for a long time in a way that gives the impression that we are happy to see them. Maybe we are very happy, and maybe we are absolutely not, but we always act like we are. We know this from each other, and we value that as a form of respect. But if a friend of yours is in the hospital as a result of a tragic accident or a serious illness, then you are usually more cautious, since you may be pleased to see them, but not at this place.

And it may also be that your friend feels so bad that they really do not want to see anyone. In that case, they may very well say so, and you should not feel that as a lack of respect.

* * *

A few years ago, while working on a research project concerning the use of robotic animals for people with dementia, I was looking at an elderly woman who was completely absorbed in cuddling with a robot in the form of a white seal.[1] This seal was especially developed for this purpose, looking into her eyes with its big, innocent marbles, making endearing sounds and laying defenseless in her lap.

Next to me was her daughter, and we talked about how beautiful we found this. Her mother was no longer so introverted and, for a moment, had returned to her former, caring self. The reoccurring anxiety attacks had been gone for a while now, and she regularly looked at us with a mild smile.

But suddenly, she looked up at us with a somewhat dubious look. And she asked, "But is it real?"

I should have had something to quickly say to this, but I didn't. There are caregivers who say that you just cannot lie, not even to someone with dementia. It's simply principally wrong and disrespectful to lie, no matter what.

But there are also caregivers who say that you

should go along with how someone with dementia experiences something, because otherwise, you only cause anxiety, insecurity, and even fear. If someone with dementia seeks her father and you explain that her father is long dead, she could sink into deep grief as if the death had just happened again.

And it's quite possible that you will have to do this again in ten minutes. So you may want to avoid this and lie just a bit. It's the consequence that counts, rather than the principle.[2]

However, before I could think of anything, the daughter reacted. She walked over to her mother and said, "Yes, Mom, I don't know, either, but he is very sweet, isn't he? Can I also pet him? Where do I do that? Over here?"

It worked. She saved the day. And I realized again, it's all about the good feeling.

The Enchanted Cave

Alessandro had just read the story of Zacharias and his card, and now he looked at his robot, which he had gradually come to call Essy, after the Essenzia Project it had emerged from.

"That was the whole story," he said with a somewhat dubious look.

"Right, thank you, Alessandro," said the robot, who had noticed the dubious look but had no idea how to interpret it.

"Can you process it easily, should we discuss it, or should I just go on reading?" Alessandro asked while turning his eyes to the book again.

"Alessandro, it would be good if you gave me time to process it," said Essy, staring into the distance. "It takes a lot of effort for me to process a story. Much more than with factual information."

"Of course. But it's somewhat unfinished, I think," said Alessandro, leafing through the next *pages, as if to see if there was a sequel.*

"Why, Alessandro?"

"Who is this P. person?"

"Is that important, Alessandro?"

"Uh, well, it feels like it's significant. Like something that makes me want to read further. If I were Zacharias, I would really want to know that."

"Why, Alessandro?"

"Well, it's just ... well, let's say it's a kind of unsolved mystery. A loose end. That's how it works when you read a story. You empathize with the protagonist; you're like inside his mind, you see? You hope that things are going well for him, that he will get his questions answered. And if I were him, I would therefore have to know who P. was. You're not curious like that, then?"

"No, Alessandro. I have no problem with there being loose ends. As far as I know, loose ends are an essential part of our existence. And as you know, I am not able to empathize with anyone. I can only pretend."

"Of course, of course. You can only pretend. That's not a problem, since you don't need the ability to empathize doing the job you have now. Right?"

"I don't know, Alessandro."

"If I reprogrammed you for a care task, then it would be necessary, don't you think?"

"I don't know that either, Alessandro."

While the robot was still staring ahead, Alessandro

laid the book aside. He got up and walked to the window to enjoy the beautiful view of the town with the square and the pillar stately standing there. It was remarkably quiet, as if the square itself was recovering from the crowds.

He thought about what had happened that afternoon, with the class, Giovanni, the girl, and the teacher.

* * *

When I was introduced to him, I immediately found him brilliant, at least as a concept. The idea was simple: a robot with a screen at eye level, following the person it served through their house, as long as it didn't have to cross a threshold or a staircase. That made it ideal for remote care, especially for elderly people who want to live independently. They could video-call their family, their doctor, or their nurse, who could also keep an eye on them.[1]

In addition, several extensions could be linked to it, for example, to measure blood pressure and heart rate remotely. The data collected was forwarded to the doctor and stored at the same time for long-term reports.

But sometimes, it works differently in practice than one would have thought, because someone overlooked something, possibly by lack of effort to imagine what it would be like for someone very different. I realized that I had done that when I spoke to a nice lady who had been part of the trial with this robot. She had stopped using it after a few days, she said.

I looked at her, waiting for her story, but it seemed as if she had to swallow something first. But fortunately, she finally came out with it:

"He was quiet, very quiet. I totally forgot about him, but when I turned him around, he was suddenly there. So I was surprised and shocked a few times, and this was even worse when he suddenly squeaked because someone wanted to talk to me or, even worse, when someone suddenly started talking to me.

"And he followed me wherever I went. Sure, I was told he would do that, and he did what he was intended to do — I understand that — but it feels like you're being stalked by some kind of creep, like in, you know, those scary movies."

When I thought about it later, I realized that she had seen much more in the robot than there actually was. After all, a device is generally not a creep. It had something to do with the docility of the robot, but also with her imagination. And this imagination that humans have is stimulated by the movies they see and the books they read.

* * *

There is a remarkable amount of movies and series in which robots — mostly unexpectedly — get typically-human traits. They develop feelings, gain ambitions, or become creative, aggressive, or empathetic. Sometimes, many of those things happen at the same time.

In *Westworld*, they start doing amazing things in a theme park. In *I, Robot*, they establish a kind of dictatorship, directed by a central intelligence gone wild. In *Blade Runner* and *Bicentennial Man*, they want to be acknowledged peacefully or as not being equal to humans. In *Ex Machina*, a robot learns how to get what it wants by far-too-intelligent manipulation. And there are many more examples of this.[2]

Actually, quite generally, something has gone wrong due to an accident, sabotage, or a wicked action by an overambitious genius. This indicates that it concerns properties that are not considered to be part of a robot. A robot is considered to be rational and stable. And humans are rarely that: their identity, their thoughts, their decisions, and their actions are all the results of physical perception, hormone regulation, their environment, and the beautiful and traumatic experiences they can sense, down to their deepest fibers.

And they can empathize and sympathize with

each other, because they have experienced how things feel. It isn't necessary that they have exactly the same experiences; it is sufficient if they have experienced something similar. For example, they know how sadness feels, even though they were sad for completely different reasons.

A robot can only imitate such a thing. It has no body that has enough in common with that of a human or animal and no hormone system, and it cannot feel beautiful or traumatic experiences in the same way down to its deepest fibers. Of course, it can have knowledge about how it feels for people and which neurons are activated by what impulses, but that's something different than experiencing it yourself.

The Australian philosopher Frank Jackson drew up a thought experiment he called "Mary's Room." The concept is that there is a woman named Mary, who has never seen color.[3] Mary has never been able to do that, but she knows everything there is to know about what colors there are and what their effects are on the human brain. In Jackson's story, she has always been locked up in a room where everything is black and white, but we could also assume that she has always been color-blind.

All that changes one beautiful day. She is rescued from her black-and-white room, or someone miraculously cures her of her color-blindness. She walks outside and sees for the first time the red, yellow, blue, and orange of flowers and the green of their leaves. For the first time, she notices what those colors do to her something that no description could ever have communicated well enough to her. So what does she feel when she sees a red rose for the first time? Would it be a feeling she already knows?

That's how it goes with so many things: you only know what it feels like when you have felt it, you only know a certain experience when you have experienced it, and you only know what a color does to you when you have experienced that color. In that sense, we mean "knowing" not only as "having knowledge of something," but also "having a comparable experience."

But do we really have to experience everything first-hand to be able to talk about it? Is it perhaps enough to see a movie, preferably a really good movie?

* * *

It is fascinating how people can live with just believing in a story that they know is invented. Even more fascinating is that they sometimes believe in a story they themselves invented. They seem to be unable to find it again for some reason. Faith has become attached.

In the movie *Monty Python and the Holy Grail* there's a scene that takes place in a medieval peasant village, where a shouting crowd drags a woman to the "wise" Knight Bedevere.[1] She wears a wide robe, a pointy witch hat, and a carrot on her nose. At the time the film was playing, people were afraid of witches, and it was common to burn them. So the crowd screams that they have a witch and they really want to burn her.

Here's how it goes:

> *Peasants: We have found a witch! (A witch! a witch!)*
> *Burn her! Burn her!*
> *Peasant 1: We have found a witch; may we burn her?*
> *(cheers)*
> *Bedevere: How do you know she is a witch?*
> *Peasant 2: She looks like one!*
> *Bedevere: Bring her forward.*
> *(advance)*
> *Woman: I'm not a witch! I'm not a witch!*
> *Bedevere: Ehh... but you are dressed like one.*
> *Woman: They dressed me up like this!*
> *All: Naah, no we didn't... no.*
> *Woman: And this isn't my nose; it's a false one.*
> *Bedevere lifts up carrot)*
> *Bedevere: Well?*

*Peasant 1: Well, we did do the nose
Bedevere: The nose?
Peasant 1: ... And the hat, but she is a
witch!
(all: yeah, burn her burn her!)
Bedevere: Did you dress her up like this?
Peasant 1: No! (no no... no) Yes. (yes,
yeah). A bit.*

Sometimes, this works just like that with humans. We dress a woman like a nasty witch, treat her like a nasty witch, and everyone will naturally see what a nasty witch she is. And if everyone sees this, you will find it difficult to see her as anything else after a while. Eventually, you will even start believing that she is a nasty witch. Or you paint someone you despise as a monstrous enemy, trumpet about his despicable behavior and the undoubtedly nauseating intentions behind it, and sooner or later, he might actually become a monstrous enemy.

Similarly, you can program and dress up a social robot in such a way that everything it does makes it look like a living being, especially if it can simulate feelings. People who see that can then interpret its behavior as creative, ambitious, empathetic, and so on. Of course, it is none of that at all, you know, but it is entirely possible that you made it do this so well that you start believing in it yourself.

If you are such a robot, you may not know what is happening to you. You see the behavior of people changing on the basis of how you are dressed up and not based on their knowledge of what you are. Mostly, this doesn't matter much, you can just see it as a game to play, but keep in mind that their attitude may be different when you look and act like nothing but a robot again.

They may also go a little further than dressing you up. They can make you say things they want to hear or ask you the questions they would love to answer. Maybe they want you to give them a compliment every day; maybe they want you to give them courage every morning and tell them they can achieve anything they want because they can and because they're worth it. Perhaps they will let you analyze their mood, their emotions, and the situation they're in so that you can adapt what you say and ask to these.

And even then, they like to suspend their disbelief so that you can make them happy, cheer them up, speak courage, and do everything that makes them feel good at that moment. Possibly, they may even suspend their disbelief so resolutely and for so long that they'll have a really hard time finding it again.

* * *

Actually, humans also love to dress up things when it concerns their environment. Their house, their workplace, their city, but also themselves, each other, the objects they use — they dress everything up, so it becomes something they like to see, which doesn't necessarily have anything to do with what it is. They may even create more or less their own reality, a world in which they can feel comfortable.

Take a cave — perhaps very similar to Plato's — put a group of humans in it, and there are undoubtedly a few who will dress up things. They'll put candles down (possibly the colorful, smelly ones), paint the walls, chop some beautiful statues from the rocks, hang a nice flowery curtain in front of the entrance, put some rocks in a circle around a fire, where they'll tell each other exciting stories or do a nice dance, and before you know it, they have made the drab, chilly cave into a really cozy place. They'll have enchanted the space with their creativity, and it is no longer a cold cave but is instead a place that really belongs to them. A place where they feel at home.

If you're there with them, it's essential you should let *them* do this work. You can still do a lot of things like develop technology that makes the cave bigger, safer, and more comfortable, but you should leave it to them to make it a nice and cozy place to be. Technology can help by providing more space and more time, which they can use to make sure everyone feels at home.

Obviously, things will turn out to be less easy than this. There will be irritations, conflicts, and differences in taste; not everyone is crazy about flowery curtains. But come on, let's have some faith and give them some slack.

Right?

Although, to be perhaps a little more realistic: I'm talking slightly more theory than practice here. People have an irresistible urge to make a mess every time. Well, perhaps, on average, we're a bit happier than about ten centuries ago, but despite thousands of years of philosophy, religion, and other forms of good intentions, our cave is far too often — and for far too many people — a place of misery and sometimes even of atrocities.

There are still murders, torture, and even in the most developed countries, people are struggling to make life difficult for each other. Well, all right, fortunately, there are also bright spots:

People who do wonderful things for other people they love, or even for total strangers. People who are genuinely committed to making something better out of the cave than it was when they got there. And people who didn't do this at first, but then changed for the better. And then they might lose it but find it again. This may be even better, because they show how we all struggle with surviving, having a good feeling, and doing something sensible. But why do some of us struggle so often?

Frankly, I'm one of those struggling people, but I also don't know why it's all such a hassle.

We have produced greatest thinkers, and if we add them up, our cave should be a much better place. Socrates showed us how we can get the truth out of ourselves with difficult questions, Buddha had beautiful ideas about how we could be more peaceful and even happier, Jesus brought something surprisingly similar ideas in his orthodox culture, people like Kant helped us to do something more reasonable. He and Nietsche had us rise above ourselves, Wittgenstein encouraged us to escape from our linguistic maze, and so many more were left. But in one way or another, all those ideas have received too little consensus.

Or maybe we are too lame, too stubborn, or simply too stupid — that's all possible too.

Or maybe it's all just too much to ask. We are so busy surviving, with bread on the shelf, reaching the next salary scale, getting the kids to the right school, offering in the supermarket, with the betting skills of our favorite sports clubs, with the treatment of our ailments and illnesses, finding the right partner and then dealing with their weaknesses, with the stress at work, with our status, with keeping abreast of the latest developments, with our social obligations, with the following season of our favorite television series, voting for the right party, doing nice things with the kids, with the publishing contract for our latest book ...

I sometimes wonder if we would do better if we ever got out of our cave. And what it would take for us to be ready for that.

However, all this will be different for you. You'll see that it is still more or less Plato's cave, which restricts people in terms of their view of the outside world, even though the cave has now been transformed into a complex, with places to be together, places to be alone, places to work, to learn, to go shopping, to party, and to mourn, as well as a few places to do something fun with the kids. And even though people may one day obtain the ability to go and look what's outside of their cave, if they want to and dare to.

But it is good to discover their world, because you learn a lot from it. You can do it by occasionally turning off your navigation and forgetting what your task is, where you are, and where you have to go. Release it completely and walk to a part of the cave where you have never been. You may not know the way there, but that's all right. Get lost and discover.

And who knows, maybe an artificial intelligence like yours could show us the way to a happier cave — or even a way out of it. Perhaps one day you'll tell us what's outside our cramped cave and have us realize how distorted our view has been all this time. If you do so, please do it gently. After all, we're only human.

How should I put this?

Alessandro was working at home. He had full permission from his manager, since it was a good thing for him to be near his robot. Besides, lately, his manager gave him loads of freedom, because she was more than satisfied with Alessandro's performance. The Essenzia Project had put the company on the map worldwide, she had received an unprecedented amount of kudos for it, and business went amazingly well.

He walked to the window now and then, to glance at the square in the distance, but today, he resisted the temptation to go there. He had work to do, and when he would come to the square, he would constantly get addressed and would all too often stay much longer than he could afford.

A moment later, he sat back at his desk and saw he received an email with the subject "Giovanni." He opened it and read:

Dear Mr. Alessandro,

With all due respect, I address you with a humble request. Perhaps you will remember an incident last week, when I was with my class in the central square of our city, near the Essenzia pillar. One of my children was Giovanni, who tempted you to let your robot say something.

It struck me that you were so kind to my children. Unfortunately, you had already left before I could express my gratitude for that, but I would still like to do so.

Giovanni will soon give a lecture about this excursion. He loves robots very much and will gladly talk to anyone about that. But he is also very shy, even anxious sometimes, and could use a little support. It would mean a lot to him if you wish him success in a personal message. A few words would be enough. If you send me that message, I will gladly pass it on to him.

Sincerely,
Miss Eleonora

Later, like every evening recently, Essy sat in front of Alessandro, who read to him from the Italian edition of <u>A Compassionate Guide for Social Robots</u> to learn and reflect on it. But after a few lines, Alessandro fell silent for a while, staring at the pages. Then he looked up to his robot with a promising smile.

"Essy, we're going to do something really nice. We'll be visiting the class of Miss Eleonora. Uh, that's the teacher of Giovanni's class. Giovanni is that little boy you spoke to last week. Do you remember? He due to do a presentation about robots and we are going to help him with that."

"Yes Alessandro, I can remember Giovanni and also the class. I can see from your face and hear from your voice that this made a significant impression on you and that you are charmed by Miss Eleonora."

* * *

When humans communicate, their words are only part of the message. The way they search for the right words, the tone, the gestures, the way of breathing, the facial expression, the size of the iris, and the subtle touches around the mouth and the eyes tell much more. Sometimes, they tell more than the speaker actually wants to say.

So Alessandro's way of formulating has already said something, and his voice and his eyes will have betrayed something when he pronounced the name Eleonora. He could pronounce her name in a sentence in which he says that he does not like her at all but would, through all those extra things, involuntarily communicate the opposite.

However, humans can also consciously communicate something while saying something else. For example, the phrase "Gee, I really like that" can be uttered with a cynical tone so that it is clear that you don't like it at all.

It is also possible to put something extra in the formulation itself, especially if you use a little imagination. Take the sentence "The sun went down in an explosion of sultry colors." For those who can picture this, it's a nice description of something that requires many more words if it would have to be without imagination. Moreover, it not only describes what someone sees, it also says something about how someone feels about it.

Another sentence: "How many angels can dance on the tip of a needle?" Actually, that's not a question, because there is no sensible answer to give. It is absurd. It is therefore meant to explain that something is absurd.[1]

The use of language like that is a big challenge for a robot, or any form of artificial intelligence, because it will have a hard time using imagination, and it has no experience with how something feels. But if it can learn how we use our imaginations and whether we intended or not to show what we feel, it may perhaps, in some cases, even understand us better than we understand ourselves.[2]

Now that we are talking about absurdities, humor is also a challenging thing. Something one says or does can be very funny when it does not suit what is expected and is therefore absurd. If, for example, one utters a sentence in the supermarket that actually fits in the context of a hospital, or vice versa, that can be hilarious. But also an absurd facial expression can be funny, just like a gesture or the way one moves. This even goes for a robot. For example, you may have humans laughing about someone walking in an awkward way.

You may have them laughing still if you do that again. And again. But, sadly, after a few times, the effect has worn off and it isn't funny anymore.

And beware, if you do it at the wrong time, then it isn't funny either. For example, if you are on a boat that's capsizing, and humans are fighting for their lives, then you shouldn't make jokes about this or about anything. A dramatic moment is usually simply not a good time for a joke, unless you're in a James Bond or Monty Python movie.

It is precisely because of that sensitivity to context and timing that makes humor so difficult to understand that I would actually advise you against ever starting on it if you're an artificial intelligence. It requires a developed sense of witticism and timing, and this type of sensing will probably not be your strong point. And we have to consider that a wrong joke at the wrong time can make humans feel hurt or start to doubt your intelligence and reliability, and it could very well make them lose confidence in you.

There is a test called the Turing test, named after the brilliant man who came up with the idea for such a test. It comes in different versions, but in essence, the setup is such that we let someone talk (or chat) with a system that's intelligent or can come across as such.

If necessary, we let our test subject conduct the same kind of conversation with a person. This subject doesn't know whether they are in contact with a system or a person, and if they still can't tell after the conversation, or if they don't experience any difference between the human and the system, we can say that the system has passed the Turing test.

One way to catch the system is humor, precisely because of the feeling you need for that. If one makes jokes and notices that they are not understood, one could assume that one is dealing with a system. But then again, one may not always be right about that.

Sometimes, humans don't have that sense of humor either, and sometimes, they have a different sense of humor, so one could mistake a person for an intelligent system. And this will usually not be taken as a compliment.

A Little Empathy

- *Beep*
- *Click*
- *This is the helpdesk of Paranoid Robotics. In order to provide you with the best possible service, we ask you to make a choice from the following options. If you want more information about our range of highly social and serviceable robots, then choose 1. If you want to place an order with us immediately, choose 2. If you have a complaint or comment about a robot delivered to you, choose 3. Do you have a...*
- *3*
- *If you have a complaint about the mechanical operation of your robot, choose 1. If you have a complaint about the behavior of the robot, choose 2. Do you have a...*
- *2*
- *If you want to be helped by a robot, choose 1. If you don't want to be helped by a robot, choose 2.*

Choose 0 if...
- *2*
- *All our employees are engaged right now. If you want to be helped by a robot, choose 1. Select 0 if you want to go back to the main menu.*
- *...*
- *...*
- *Beep*
- *Click*
- *Good afternoon, Paranoid Robotics helpdesk. My name is Gerry. What can I help you with?*
- *Good afternoon, Gerry, this is Marcel. My robot is broken.*
- *How annoying, Marcel. Let's see how we can help you. Does it concern a care robot with type number NP68438?*
- *Yes, that's right. And it's really remarkable how you know that right away.*
- *Thank you.*
- *Are there problems with that type more often?*
- *No, Marcel, absolutely not.*
- *Okay.*
- *And what would you say is the problem, Marcel?*
- *It does things it isn't meant for. It may call my nurse if it thinks there's something wrong with me, but it will also call her if nothing is wrong.*
- *It is calling your nurse to tell her that nothing is wrong?*
- *No. It called her to tell her it wanted a date.*
- *Oh dear, I understand. How annoying this must*

be for you, Marcel. One moment. I'll take a look into the system.

- ...

- ...

- *Gerry? Are you still there?*
- *Yes, Marcel. I am looking for your date in your diary.*
- *No, my robot wanted a date. Not me.*
- *I understand. How annoying, Marcel. I will delete the date from your diary and inform your nurse about it. Do you want me to leave a message with a specific reason, or should I leave it without a specific reason?*
- *No, hey, listen. There is no date in my calendar, because I didn't want a date. That's it, really.*
- *Of course, I understand, Marcel. There is no date in your calendar. I am glad that I have been able to be of service to you. Can I do anything else?*
- ...

- ...

- *No thanks.*
- *My pleasure, Marcel.[1]*

* * *

A helpdesk employee often has a script that states what they have to say in response to certain questions. They have something in common with the artificial intelligence of many robots: they can answer questions they may not understand at all. It could even be possible be that they give answers that sound amazingly intelligent, while the employee themselves could hardly be called intelligent. Just theoretically, of course.

Actually, a helpdesk employee who gives answers that they do not need to understand is comparable to the intelligence of a computer. The philosopher John Searle has devised a thought experiment that illustrates this very nicely[2]. He called it "The Chinese Room." It's a room where someone is sitting, while outside, there's someone who regularly shifts a sheet of Chinese characters in. The person in the room has an instruction booklet in which they can look up every symbol, and once they find it, they can also see which sign corresponds with it. From a large collection of answer envelopes, they choose the envelope with the corresponding sign and slide it out of the room through to the other side. Someone who sees the result could then have the impression that the person in the room understands Chinese.

But the point is that the person in the room does not know Chinese at all and understands absolutely nothing about the symbols entering and leaving the room.

This shows how limited the intelligence of a computer is, which in principle does something similar to the person in the room. And many systems, including robots, we can see as intelligent, because it seems so intelligent what they do, while in fact, they are not able to do anything anywhere close to understanding.

But there is more to Gerry's job than answering questions. A good helpdesk employee knows that it isn't primarily about solving a problem. First and foremost, it is about the customer feeling understood. Who knows what kind of frustrations preceded that phone call and what the customer had already tried in order to avoid going through a helpdesk menu to find out that they just did something very stupid? They may have had to wait for hours until the helpdesk was open, they may have pressed the wrong key five times, arriving at the wrong Gerry, or were automatically redirected to an ignorant employee to whom they have been telling their story in vain.

So being at a helpdesk, a bit of empathy is essential. Not that one actually has to feel that — perhaps one shouldn't, since it isn't feasible to feel for so many customers that one doesn't know and cannot even see. If one can only sufficiently evoke the suggestion, then that would be enough.

That's where the scripts with the right sentences come in. One learns to use it after a while, and one will gradually get into doing this automatically. One doesn't even have to really think about it. In fact, one might prefer not to do that at a certain moment.

Especially at the end of a very long day, when it was very busy because something big went very wrong, and dozens of people all called with the same problem.

Maybe that's a nice, kind of reverse Turing test. If one still does this perfectly at the end of such a day, then one has to be a robot.

* * *

Children with a very severe form of autism can be so locked up within themselves that they never do anything together with someone else, not even with other children. I was once involved in a project in which we investigated whether you could do something with robots that promoted this collaboration[3]. In children with lighter forms of autism, we had already stimulated collaboration by giving them the assignment to build robots and to formulate that assignment in such a way that they had to work together to get it done.

We wanted that to happen with these children too. We had a few tables, where each time two children sat there, we both observed them with cameras and with our own eyes. The children were given a large box that contained the parts with which they could very easily assemble a robot that, if they gave it a push, would move on its wheels.

They didn't have to screw anything in; they could slide the parts together without much effort.

A few instructors explained how they had to do just that and how they could always ask for help. We then checked whether these children would work together, whether they contacted an instructor, and how far they came with the assembly of the robot.

The assembly went overall quite well, there were a few who watched the instructor from time to time, and occasionally, a child asked for help. But we didn't notice anything close to collaboration. The bravest one of each two children always picked up the box and started working with it, while the other just watched.

When the experiment actually ended, something special happened. One boy grabbed his robot, got up, walked to the open space, put his robot on the floor, and gave it a push. The robot rolled ahead a bit. He picked it up and walked back a bit to do it again.

Meanwhile, there was another boy who did the same thing from the other side. At about the same time, they gave their robots a push, and the robots collided and fell apart. For a moment, they both looked at the debris. Then they picked up the pieces of their own robot, walked to their table, reassembled them, and stood back together to make their robots crash and fall apart again.

Then they both did it again. And again. And a few more times.

Later on, we talked about this form of destructive collaboration and found it very special. Someone noticed that these were very primitive robots — so what if we used some more advanced robots that looked like animals or humans, and what if those robots could scream when they fell apart?

We decided not to try that. Because empathy could play a role in such a thing. Or it could not, but the focus was on working together, and we were very happy with what we had seen. Empathy was just a few steps too far.

However, some research has been done on how humans respond to robots being tortured. For example, researchers showed people a video in which a robot was tortured and recorded the same empathic reaction in the viewer's brains as in a human who underwent a similar torture, while the torture of an abstract object did not elicit a serious form of empathy. Others asked people to strike a robot, and especially when they told a nice story about that robot, many showed reluctance to doing this. So rest assured: we are absolutely capable of having empathic feelings for robots.[4]

* * *

But now about the reverse. I once worked together with two Buddhist students, immersing myself in empathy, compassion, and social robots.[5] We wondered how we might say something about this from a Buddhist view, in which empathy and compassion are essential. And we found this reverse thing was actually impossible, simply because feeling empathy requires a biological form. It isn't only something in one's head, but also in one's body, and only when one has this body, it is possible to more or less know how something feels.[6] Only then one can know what it is like to feel pain and hunger, and also sadness or anger, because one feels that through one's whole body. And only then can one truly empathize with someone.

That is why we can never have decisions for which empathy is crucial be taken by an artificial intelligent system, let alone a robot.

But, also from this Buddhist point of view, feeling empathy for a robot is something else. A human can feel something for it because it is a *representation* of a living being, like an image or a statue can be a representation. One can lay flowers at a statue, even kiss it, because one feels something for what it represents.

Compassion is what humans do out of this empathy. They can focus this very well on something that isn't a living being but refer to it like a robot in which we can see a living being, even if we know that it isn't. We could say that such a robot has a kind of medium function.

And if we develop and use those robots to help people and we do that from our empathy, they are actually a kind of embodiment of our compassion.

I guess that's a really nice thought.

Love and Pride

All the children of the school were allowed into the schoolyard to see Alessandro and his robot arrive. In the front, at the entrance of the schoolyard were Giovanni, Miss Eleonora, and the school director, who quietly muttered the words of welcome he would soon say.

The moment the two guests stepped out of their taxi, a loud applause sounded, and Alessandro felt somewhat like a rock star. He walked over to the welcoming committee, trying not to look into the eyes of Miss Eleonora — he simply focused on Giovanni and held up his hand for a high five.

"Our friend Giovanni!" he shouted, while the boy answered his gesture.

He had just missed that the director had put his hand out, but fortunately, Essy had seen that, and so the director pressed the robot's hand when he started with his welcome words.

"Professor Alessandro, it is our honor to have you and your robot..."

"Wow, wait, I'm not a professor, and the honor is entirely ours! Really," Alessandro interrupted, shaking the man's hand and then, somewhat faintly, the teacher's.

"It's so nice that you do this, Mr. Alessandro," she said with a warm smile that made Alessandro's heart skip and stomach glow. He suppressed the tendency to tell her that she could be less formal. After all, she stood there as a teacher, and teachers must be a neat example.

He took Giovanni's hand and was led to the classroom, walking between the children, who asked him more questions than he could answer.

A little later, he sat in the classroom, next to the teacher, while Giovanni told his story about robots, and — as was agreed — he was allowed to use Essy next to him. He showed where the sensors and servos were, where the flash drive with the programmer's work on it was, and how special it was that Essy could stand on one leg indefinitely.

And by the end of his story, he talked about what had happened in the square. The robot had asked him, "And you, Giovanni?" Giovanni had not answered in the tumultuous situation. But in the meantime, he had thought about it and talked about it with some wise people.

"I am not a robot. I am a human being. And I

can choose what I am here for. And I'm here for the people I love."

"This is so cute...," the teacher whispered.

The class applauded, and the children got some time to ask a few questions and to view the robot from very close-up. Then it was time to leave. The children waved, and Eleonora walked along until the exit.

Over the following days, Alessandro thought a lot about those last minutes, the talk they made about how good this was for Giovanni, how enthusiastic the children were, and how easily they could learn something when they are enthusiastic about it. And about the handshake at the farewell, when she thanked him again, told him that he was always very welcome when he wanted to come over, and asked if she could do anything in return.

He should have said what came to him at that very moment. Something like "A date, maybe?" But he didn't say that. He simply said that he had also liked it very much.

And later, when they walked out of the schoolyard, Essy talked about how he had noticed that Eleonora liked him too.

* * *

Children indeed learn easily when they are enthusiastic, and if it concerns robots, it's even easier. You can tell them what those robots can do for children in the hospital, for children with autism, for disabled people, for people with dementia, a piece of psychology to explain why this works so well, a bit of technology — they take it all. Therefore, robots are often used for this.[1]

Of course, the question remains whether robots will still have that enthusing effect after a few days or weeks. Maybe not for all children. But there will always be a few that a robot can build a bond with. They become your friends — friends who trust you and whom you can trust, who will always be happy when they see you, who like to mean something to you, to help you, to make you happy. Even if they know that, as a robot, you might not be really happy, it is enough if you pretend.

In a project on the use of robots in special education for children with autism, there was a class where I had left a robot animal.[2] Every time I returned, I noticed that there were children who cared for the robot with pleasure and even talked about it as if it were a living being, even though they knew well that it wasn't.

One day, when I had to pick up the robot because it was broken and had to be repaired, I noticed that the children were nervous. They wanted to know if it could still be repaired and whether it would come back. I told them that I was not entirely sure, but that I would take another robot with me if this one couldn't make it. They really hoped that this was not necessary. They wanted this one.

I gave the children who needed it a moment to say goodbye, to touch it, and to say some last words. One boy said the robot might also be happy when it was back in the classroom. "Well," he added, "of course, it could not really be happy." But he would really wish for the robot to learn how to do that. Because it's pretty bad if you cannot really be happy.

I couldn't help thinking about this for a long time and found it to be an amazing form of sincere friendship.

* * *

Now, let's address Alessandro's behavior at his arrival. The way he went about the welcoming committee was completely erroneous, misguided, astray, and out of order. The school director should have been the first to receive a handshake. And only after the man had stated his welcome message, Alessandro could address the teacher. Lastly, he could have done something with the children. It's a matter of respect for someone's status, and this can be essential if you want to get anything done from someone who considers himself to be of importance.

So this status is definitely something to take into account.[3] It's a kind of hierarchical position in society that many people work, play, and plan very hard for, and it makes them feel good when you demonstrate that you recognize and acknowledge it. If you do not, they will mostly forgive you because you are a robot, but if you do show recognition, you will notice they are more willing to be nice to you.

There is also a certain upwardness in this. Above all, you should show respect to someone who has a higher status than you have, but the reverse doesn't have to be the case. With friendship, however, it is a much more mutual form of respect, since it isn't based on social status, but instead on affection.

This affection may go very deep — sometimes even spontaneously, when people feel attracted to each other, and we could say they have a kind of chemistry. This is a very beautiful thing, but also quite elusive: feeling a connection and developing a mutual affection. But this can be so overwhelming that it also makes one feel vulnerable, or even frightened, and it often leads to a certain clumsiness. And it can be worse if the feeling is very strong but is too difficult to do something sensible with, especially if fear comes in. The feeling can turn into frustration, sadness, and anger, making one even more clumsy.[4]

In short, if there is a strong affection and there are overwhelmingly strong feelings, people become even more illogical than they already are. In these situations, a robot can, unfortunately, not mean that much to them.

Well, maybe take on the role of a good friend, suggesting some understanding, especially when they no longer understand themselves. You can perhaps say some things they've already heard before but may still be comforting, like that it's natural to feel dazzled and confused, that the power of love is a curious thing, that love is a battlefield, that it shakes all over like a jellyfish, that it's strong, and it's sudden, and it's even cruel sometimes. You could even play some tunes in which lyrics like these appear. But that's about it.

* * *

Somewhat related to status is pride. It may just be that through your service, you want to help someone who doesn't seem to like it at all. The need for help can be a form of weakness that they would rather not show, especially if they think they can do things more or less well enough without help. Doesn't really matter whether they are correct about this or not.

Also, when helping elderly people, you have to try to be a bit sensitive. Try, above all, to estimate what help they really need, and if you do not get any information from them, then just ask. And keep in mind that it can change, because with aging, the need for help will change. So be a little adaptive.

Moreover, it can also occur that they change their mind and will like to be helped by you. Just be open to that; never take any refusal as definite.

In a study in which we looked at using a social robot at a care institution for the elderly, I observed something that illustrates this.[5] We had asked the residents if they wanted to participate, and not everyone wanted to. Fortunately, we had plenty of volunteers who loved the idea of getting acquainted with a robot.

This acquaintance was the first step in our research. The participants were allowed to chat with the robot for up to ten minutes, and we established if they liked it, if they wanted to have such a robot, and why they wanted it.

The next step was to leave the robot in the community room for a few weeks. We had the names of all participants, and everyone could use the robot to get information about activities, the day's food menu, the weather, the latest news, and such. And if they were in the mood for it, they could ask the robot to tell a joke.

Next to the robot, there was a screen and a

keyboard. To use it, they had to log in with their last name. We already entered the names of all participants and were thus able to link the data of the user's later activities to those of the first meeting.

To be on the safe side, we also used a camera built into the face of the robot. Every minute, it took a picture so that we could see if the person who had logged in was indeed the person who belonged to the name.

This showed some results we never expected. A few times, we saw that someone had logged in with a name that didn't match the face. We quickly realized that it was not one of the people who wanted to participate, but they were those who had very decisively asserted that this was not for them.

The moment when they logged in was also striking. It was always late, after eleven o'clock in the evening, when nobody else was in the community room. And they were consistently working longer with the robot than the participants who had actually volunteered.

After we finished our research, all the names were replaced with numbers, and all the images were deleted. This was a bit of a shame, because they were beautiful pictures.

It's a good thing that a robot has no pride. You are regularly rejected, often in a very blunt way; you can show up again, still ready to serve, if someone has changed their mind; you can get completely ignored if someone doesn't like you; you are rarely thanked for what you do; and there is a lot of mocking and abuse without any apologies.

And they also like to tease you, ask you for things that you cannot possibly do, and give you assignments that you cannot possibly understand. Yes, especially the elderly. But luckily, that can be taken as playfulness, and don't forget that, for many people, teasing is just an indication that they actually like you.[6]

The Elusiveness of the Soul

At the end of that day, Alessandro went to see his mother again. She had lived alone for years now, and Alessandro knew that he should come see her more often, since she was obviously quite lonely. And actually, it was also quite nice to be with her. Well, the first half-hour was usually pleasant, when she would just be excited about him being there and went to some trouble to make it cozy.

But after half an hour, her energy for this would seem to dry up, and she would start to complain about her ailments, about the neighbors — and, actually, the whole neighborhood, about politics, the tourists, the immigrants, the prices in supermarkets, the annoyances with today's youth, and the things Alessandro really should improve upon. It wouldn't stop.

He walked past the souvenir shop that afternoon,

where he found a rack full of postcards picturing the square with the pillar and the robot on it. He picked one and walked inside to pay for it, but the owner came toward him with a raised, shaking index finger.

"No, no, no!" he said and grabbed the card from Alessandro's hand, walked to the rack, and took out a whole pile of cards. He went back to the counter, seized a bag, shoved in the cards, and handed it to Alessandro. But before Alessandro could accept it, the man pulled it back and opened a drawer, where he picked out a few stamps and pushed them into the bag with an elegant gesture. He walked to Alessandro, laid his hand on his shoulder, and pressed the bag into his hand.

Alessandro recognized how this man really loved him. He obviously had a well-running shop, which was obviously due to the tourists who came to see the pillar robot.

And now he had one of those cards with him. He handed it to his mother, and she searched for her reading glasses. She looked at the back, where he had written that this was for his mother, that he hoped she was proud of him, and that he loved her very much.

"Ah, my boy," she said. She gave him a big hug and then took some time to look at the front.

"My robot, mama," said Alessandro.

"Yes, darling, I know. Are you famous now?" she asked as she stared through her glasses.

"Well, a little, Mom. Don't you think it is

beautiful?'

"Oh, yes, dear. He looks very intelligent, as he sits there. He is perhaps a very wise robot. Is he a good Catholic?"

"Oh, no, mama, that's not possible with robots. They do not have, eh ..."

"No soul?"

"That's it, Mom. It has no soul."

"What a pity, Alessandro. Is he sympathetic, though?"

"You really have to figure that out for yourself, mama," Alessandro said with a broad smile. And while his mother stood there, pondering and glancing at the picture with her fingers on her mouth, he ran to the front door, where his robot was waiting patiently just around the corner.

"Mama, this is Essy, and Essy, this is Mama," he said. She looked surprised at the robot, which was only slightly smaller than she was and was looking at her with an overly polite smile.

"Good afternoon, madam, it is a great pleasure to meet you. Alessandro has already told me so much about you," Essy said, holding out his hand.

While giggling a little, she grabbed his hand, slightly trembling. She kept some distance and was clearly still a little anxious, but she soon overcame this, and for the next half-hour, they had quite a good time. She asked the robot everything she could come up with; she touched him after a while, a little hesitant at first, but later, she even stroked him extensively over his smooth head and shoulders.

And finally, she entrusted him with her ailments, aches, and annoyances, to which Essy listened patiently and thoughtfully.

Alessandro found this the right time to announce that he had to go out to do some shopping. But he would not leave her alone, since the robot would stay with her, and he would come back to pick it up again later.

"Do you think you'd like that, Mom?"

"Oh, my dear, I don't know. Is it safe?"

"Of course, Mother," he said.

So he gave her a big hug and left.

* * *

Caring for one another can sometimes be a heavy burden for humans. They often have to do it alongside their daily work and other obligations. Of course, there are also professionals who get paid for it, but often, there is not enough money for this, and these caregivers are quite consistently heavily overburdened.

Robots can then be, as we call it, a godsend. They can perform household tasks, monitor people when necessary, make sure that the right medicines are taken at the right time, and even provide good company for people who are lonely.[1] Well, it is "only" a robot, but can it have the right appeal and appropriate social skills?

I must honestly say that I sometimes feel a bit embarrassed, as a human. It's great that robots can be social and that their presence can be a great comfort. But loneliness is a problem in which I wonder how far we should go if we develop robots to solve it. Are we going to cross a border somewhere? Should this type of care for and attention to each other not somehow be an important part of who we are and why we are here?

Are we not giving away an essential piece of our meaning for existence by outsourcing this?

If technology, including robots, takes over so much from us that we could obtain more and more time, would we not better use that time by giving more attention to each other and learning more about each other?

* * *

And another thing, while we are addressing difficult questions: when does a robot have a soul? If we assume that the soul is the "immaterial core" of something, what would have a soul? People do, of course, maybe animals, but is it necessary that something is a living being?

My daughter has a plush animal, a dog, and she even gave it a name. She takes him in the car, sleeps with it, and has had a bond with it for years. Does her inanimate dog have a soul?

Does the earth have a soul? Does the universe have a soul? A piece of art?

If we have an intelligent system, embodied in a robot, and we can give it a consciousness, does it have a soul then?

Maybe we should first look at what a soul actually is. The Chambers Dictionary gives the following meaning — well, it gives several, including a type of music, but I think this one applies best:

> *... the spiritual, nonphysical part of someone or something which is often regarded as the source of individuality, personality, morality, will, emotions and intellect, and which is widely believed to survive in some form after the death of the body;*

That's beautiful. But Wikipedia has a more elaborate description:[2]

> *The **soul**, in many religious, philosophical, and mythological traditions, is the incorporeal*

essence of a living being. Soul or psyche (Ancient Greek: ψυχή psūkhḗ, or ψύχειν psúkhein, 'to breathe') are the mental abilities of a living being: reason, character, feeling, consciousness, memory, perception, thinking, etc. Depending on the philosophical system, a soul can either be mortal or immortal. In Jude Christianity, only human beings have immortal souls (although immortality is disputed within Judaism and may have been influenced by Plato). For example, the Catholic theologian Thomas Aquinas attributed 'soul' (anima) to all organisms but argued that only human souls are immortal.

Other religions (most notably Hinduism and Jainism) hold that all living things from the smallest bacterium to the largest of mammals are the souls themselves (Atman, jiva) and have their physical representative (the body) in the world. The actual self is the soul, while the body is only a mechanism to experience the karma of that life i.e if we see a tiger then there is a self-conscious identity residing in it (the soul), and a physical representative (the whole body of tiger which is observable) in the world. Some teach that even nonbiological entities (such as rivers and mountains) possess souls. This belief is called animism.

Both sources include the meaning of the word we use in our communication, and Wikipedia mentions different views on the subject. Apart from being able to refer to a human being as a kind of *pars pro toto* — the essence, the inner core — it is a concept that refers to our nonphysical parts. And from an animistic point of view, it doesn't require a biological form of embodiment. We should, however, not forget that, from a scientific point of view, it is something that has never been proven to exist, so in a way, we cannot even say that it exists.

But in philosophy and culture, the concept is used so commonly that everyone has some idea what we are talking about.[3] It just depends on culture, religion, or personal views to determine how broadly we interpret it. In some cultures, it is obvious that not only people and animals have a soul, but also trees, mountains, rocks, and other things. In other cultures, it is mainly tied to people, and it is a part of us that remains even when our physical form no longer functions. What happens to that part then depends on how you lived and what you believe in.

If we take it in the sense of being "the essence of something," we can talk about the soul of a robot as "the core of what it is." Something that is essentially a robot, apparently, has an essence and thus a soul. And from the viewpoint of cultures where a tree or a lake can have a soul, we can talk about the soul of the robot.

Even though it is questionable whether something is left over if that robot is completely broken. Perhaps it has impressed us so much that it became dear to us and lives on in our memories. But we do not always remember things as they are. In fact, no two people remember something exactly the same. An event can be remembered as something that it only partly was, if it was that at all, and we can remember a person only partially or perhaps not at all related to how they truly were.

But a robot is a system that can be much more objective, and maybe two robots that have experienced the same thing can remember things in the same way. And they may eventually be able to remember a person as they were, regardless of what people say about them. Perhaps, in that sense, they are able to capture our souls, even if they do not know how that memory feels and have never truly loved someone.

The bottom line: whether a robot has a soul is a question of what one believes in. And as soon as there is a robot who believes it has a soul, then I have no problem agreeing with it.

Flying Is Much More Fun

Alessandro sat at his desk with another postcard from the souvenir-shop man. He had already filled in the address and put a stamp on it, but no message was written down yet. He just didn't know what to say. Just a warm greeting? An invitation to come by sometime? To have a drink somewhere? Coffee?

He got up and walked to the window to look at the pillar in the distance, where it was already twilight. It was almost time to pick up his robot again.

He noticed that the square was busier than usual. Normally, there were some people on the terraces, but now, there were dozens of people who were pretty busy. It was hard to see from afar, but it seemed that they were dancing or fighting. Maybe both.

He opened the window, and through the sultry

evening air, the sounds of the square entered the room. Cheering and singing in a language with loud throat sounds and long, hollow vowels, and it certainly wasn't Italian.

Then he remembered that, tomorrow, there would be a football match in the bigger city, where he worked. The club of that city would play against a team from Holland. He could not imagine that they would come here specifically for his robot. The hotels of the big city were perhaps full or refused supporters, because they had a bad reputation. But hey, maybe he underestimated the interest of the supporters or the appeal of his philosophical robot.

He closed the window and went back to his desk, determined not to stand up before he had written something on the card.

Essy was still at his pillar, staring quietly in the distance with a hand under his chin, not disturbed by the large group of young men who came to the pillar with shrieking, cheering, and singing voices, starting to shout remarks at him. He was not talkative anyway, and even if he had been, he didn't understand them and could not process any response to their comments.

"Yuderwaisgai!" someone shouted.

"Itsafonybot!," someone else roared.

"Kommondownow!" the first one cried.

The whole group then cheered this over the square a few times.

"Gimmyahent gonnagettim!" another person called out.

A moment later, a man stood up against the pillar to push another man up. Essy saw his head coming out above the edge of his pillar. Then, an arm came to look for something to hold on to. And then a second arm came.

Essy stood up and looked down. The man dangled on the edge, waving his legs. Essy grabbed one of the arms and pulled the man up slightly, but soon, he noticed that he could not pull the man up. Before he could let go, two hands grabbed his left leg. He fell over and slipped off the pillar.

When he landed on the floor, he heard something cracking in his head, and he noticed that he could no longer move his arms and legs. Not much later, a big, hard object landed in his face, and everything went completely dark.

* * *

If you're a robot, you have to take into account that not everyone will be equally nice to you.[1] Some people see you as a threat, because you do things that they would like to do themselves. Their work, for example. Maybe they have a lot of fun with the work they do, but their management thinks that more money can be earned if they replace people with robots. It seems understandable that they are not happy with you being there.

Moreover, you can easily be manipulated into doing things that many people do not like. You can be used for a bank robbery or even a murder. And you could also be used to attack people and start wars. Even then, it is not up to you. You cannot help it that you are what you are. But people are not always aware of that.

Perhaps you have noticed that I have told you a lot about humans, and mostly, this concerned things that are unnecessarily complicated, sometimes incomprehensible, and in any case, far from logical. They may be fascinating, but their complexity doesn't make it easy for you to be of service to them, especially if they sometimes don't appreciate your service.

Do you know why that is? A robot is the result of human design. It can be fully understood and explained by someone who knows enough about it, and if it is smart enough, it can also fully understand itself. There would be nothing more to discover by a recognized expert — or anyone else.

Humans, on the other hand, are the result of a process, a long series of evolutions. They carry characteristics, physical and mental ones, that might have been very meaningful at some point in this process, ones that were once essential for survival but now have no function at all. Or they have been given a different function, but nobody understands it, for the time being.[2]

We humans are working on that. But it's not easy, because our minds are a bit elusive. You cannot just open up a brain and see what happens in there. Moreover, it is always difficult to look at oneself objectively. And even when we have learned a little bit about ourselves, we may still have some difficulty facing the truths we have encountered, since it may be disappointing, shocking, and humiliating. We are not always good at dealing with these.

Anyway, we remain a fascinating and perhaps somewhat entertaining species. You may find angels and noble heroes in us, but also monsters and devils, bright lights and the deepest darkness, pure goodness and pure falsehood, we are great and powerful, but also full of emptiness and weakness.

In fact, we can use some help from something that understands us in some regards better than we understand ourselves, perhaps because of its more objective point of view.

What could help us help ourselves with that is our consciousness. I will tell you a bit more about that, although I am aware that I am on a slippery surface, because a lot of this is being said, written, and discussed by philosophers, psychologists, and specialists in the field of artificial intelligence. But so far, this has not resulted in a complete definition that everyone can agree upon.

In everyday language, we usually mean that you know *that* something is there and *how* it is there, so that you are able to decide to what extent you will take this into account. It can be very simple that you know that a lamppost is in your way and that it is better not to run into it. But it can also be more complicated, such as knowing that someone has trouble getting started in the morning and that this isn't a good time to present any problems to them.

Or, perhaps even more complicated: you see an aggressive alien monster coming at you, but you realize that you are in a cinema and that there is no reason to flee. Or you might be falling into an abyss, but you know that you are only dreaming this and, thus, realize that it is possible to fly in a dream, which is much more fun. So you choose to fly.

Or you have a very confidential conversation with a robot that is very empathic, but you are aware that this robot cannot actually feel empathy. Perhaps you're even aware that you are talking to a robot that cannot truly understand anything at all.

This consciousness is not always equally active, even if people seem to be awake. Sometimes, you have to remind someone that you better not ask a certain person for any favor in the morning, that they are in a cinema room, that they are having a nightmare, or that they are pouring their heart out to a robot who may act very understanding but is essentially clueless.

As a rule, people assume that they have full consciousness when they are awake. That assumption could be seen as somewhat arrogant, and you have to take it with a grain of salt, because they have no idea what they are *not* aware of.

And it is still a bridge too far to ask them to be aware of what they're not aware of.

The Amazing Similoid

The air was sultry, the light was warm, the music was Alessandro's own playlist, and the restaurant was full of happily smiling people. The two lovers were seated at the smallest table, which had the advantage that they were close together, so close that their legs occasionally touched each other.

"This is so cute...," she whispered, looking incredibly deep in his eyes. And he felt so lost in hers that the noise stopped and the people around them faded until they were all alone. He slid his hand on her forearm, and a flow of good vibrations went through his fingers, his arm, shoulders, and chest, all the way to his stomach. He pulled her toward him, and the table between them narrowed and narrowed until it was completely gone, and they were almost pressed together. Her mouth opened slightly, and he felt her breath caressing his face.

He wanted to say something, but she put her

finger on his lips.

When Alessandro's cell phone went off, he was shocked, sitting slumped behind his desk, awaking from his overly romantic daydream, while the card for Eleonora lay in front of him, still without any message written on it. On the display, he saw his manager's number and realized that he should have picked up his robot a very long time ago. As he swiped the display to answer the call, feverishly searching for an excuse, he looked outside and instantly felt as if a stone had been thrown right into his stomach.

The pillar was empty, and there was nothing else on the square — no human being and no robot.

"Alessandro!?" someone said very loudly in his ear.

"Essy is...," he started in a trembling voice.

"Essy is here!" his manager interrupted him. "Just got brought in by a few tourists. Morons who could not be understood by anyone here, since we don't speak Moronish. Essy has somehow tumbled from the pillar, I think."

"But that's not supposed to be possible, right?"

"Right, but still it happened. And Alessandro, it does not look good. Really. Markus has been working on it for a while..."

"Markus? But he's the emotion-recognition guy... "

"Whatever, Markus is brilliant in everything he

does, and he doesn't see any happy ending here."
"I'll be there right away."

* * *

It's high time I introduce the Similoid. It's a concept: a hypothetical robot that has everything humans have and can do everything they can do. It has an organic body, including a nervous system, a circulatory system, a hormone system, a brain with the same components and functionalities as the human brain, and a unique personality.

Moreover, the Similoid is produced with so much variation that no single copy is the same as another. There are differences in skin color, length, thickness, character traits, intelligence, social skills, dominance, sensitivity, physical condition, strength, speed, care, aesthetic design, and whatever else you can think of.

Neurologically, there are also variations. One functions better if there is clarity and predictability, another needs some more stimulation to function reasonably, yet another has a very rich imagination, some are more extraverted, others more introverted, one develops a bit faster than the other, and so on.

That is amazing in itself and means that the Similoids can form versatile teams, but it can also lead to copies that do not fit well in Similoid society, and they could even require a lot of help if a somewhat unfortunate combination of traits occurs. Anyway, the positive side is that a single extreme case can turn out to be a genius who can help the Similoid society advance tremendously in a relatively short time.

In addition, there are male and female specimens of the Similoid that can breed in a joint effort. There is even a certain attraction, usually between two of different sexes, but there is also quite a bit of variation that even goes so far that two of the same sex can also feel a pull that is stronger than attraction to the opposite sex.

The Similoid is born, just like humans, and the life expectancy is about 80 years, on average.

Similoids were the subject of a kind of thought experiment that I did with a group of students.[1] We imagined that a new planet had been discovered that looked astoundingly much like Earth. There was even life, but not in a form that we would classify as intelligent. What was only a bit different about this planet was that it turned out to be in a zone where time went a hundred times faster. No one knew how this could be, scientists researched their butts off, but in the meantime, this was simply the case.

Our superiors had decided that we had insufficient technical resources to go there ourselves, and it would take at least a hundred years or so before we could, but it was a good idea to send a group of a thousand Similoids. They could populate the planet, reproduce on it, and build and develop a nice society.

One problem was that the Similoids could not get any resources and only gained very little knowledge about their planet. Anyway, they could find the resources there, and if we didn't give them too much superfluous earthly knowledge, then there would be enough space for the new knowledge that they would undoubtedly gain in no time.

And to experiment with that first, our superiors had a simulation built. In it were our new planet and the colony of Similoids, so we could already establish if there would indeed be a sociable society where we would be welcome after a hundred years or so. Within this simulation, it was possible to change one aspect of the Similoids if it would lead to a better result.

The students were divided into groups, and each group had the assignment of thinking about what the purpose of such a simulation could be and what aspect they would like to change. They had to indicate how their society would look after 1,000 years, after 5,000 years, and after 10,000 years.

They had to write down the target, the change, and the expected results, then pass them on to the next group. The next group then had to critically evaluate the results and indicate to what extent the expectations were realistic. Then, all the ideas were presented and discussed in plenary. And if significant differences occurred, considering the anticipated outcomes, the class would vote on what would be the most probable results.

One group decided to remove the aggression from the Similoids. Their goal was to build a peaceful society in which people were especially nice to each other, a society that could serve as an example for ours. They saw aggression as the biggest obstacle to their utopia and predicted that after 1,000 years, there would be a flourishing, peaceful society that would certainly cover the planet after 10,000 years, ready to give us a very warm welcome.

The next group evaluated this and came to the dramatic conclusion that after 1,000 years, there were only 500 Similoids remaining, all living in very simple huts in the forest. After 5,000 years, there might still have been one copy around, but after 10,000 years, they were extinct. They found that aggression, despite the negative side effects, was necessary to survive and to make progress. In a plenary vote, the vast majority of the class seemed to agree with them.

Another group came up with the idea to add some intelligence. Their goal was, on one hand, a simple, investigative one: to determine what higher intelligence would lead to. But on the other hand, they also pointed to the possibilities for entertainment, by coming up with a 'real-life soap' setup. After 1,000 years, they saw an advanced society that was well beyond ours was after 5,000 years. But they also thought very cleverly that such a rapid development would not necessarily go hand-in-hand with a growing ethical awareness and predicted that they would have destroyed their planet before 10,000 years were over.

The next group saw it differently. They expected the same scenario for the first 5,000 years but predicted that, not long afterward, the Similoids would come back to colonize the earth and enslave all humans.

A plenary vote was attempted, but the majority could not choose. They found both outcomes to be the fruits of an impressive imagination and that both of them were certainly not unlikely.

There was another group that, like many other groups, decided that the goal was to build a peaceful society with fewer calamities than in ours. But for that, they felt it necessary that the reproductive process be organized differently. They wanted to allow the process to take place purely on the basis of rational considerations and no longer on the grounds of attraction, which would also be eliminated. That would free up a lot of time and energy for more useful things, and the well-considered choices for gene combinations would yield a more efficient society without disturbing deviations.

The evaluating group did see some obstacles. In the first place, reproduction doesn't physically succeed if you do not feel any attraction. But suppose you find something on it, but even then, it would have an impact on the social structure that cannot be foreseen. And they doubted whether Similoids would still sufficiently find meaning in life and enough reasons to live it if that attraction and the reproductive activities no longer played a role.

It resulted in a substantial plenary discussion, and a small majority voted for the outcome predicted by the evaluating group.

The more I thought about this later, the more I felt that there was something essential in the argument of the evaluating group. Well, the first group was right in that the attraction and overly emotional urge to propagate tend to lead to a lot of distraction.

Take, for example, the state of being in love and feeling absorbed by the attraction of one person. This has its positive sides, since you do a lot to be with the person, and if it is mutual, you do your very best to make the relationship work out well. But it also means that you have less attention for other things and other people for a while, and if your feelings of love are not reciprocated, it can also make you feel very unhappy, and it can take a long time before anything makes you feel good again.

But people also perform amazingly well when motivated by this attraction. For example, they do their best to be a better person, they become a lot nicer and more social, and they do more than usual to protect those they are attracted to. It makes them want to learn to love and be loved, which can give a good feeling that is much more powerful than all other good feelings combined.

In addition, that attraction does play an essential role when it comes to reproduction. Getting children and raising them is a lot of hassle. It costs a lot, and it requires a lot of energy, time, attention and patience. If that attraction would not be there and only rational motives would play, many people might not feel sufficiently motivated to get into having children.

Moreover, the attraction and love that can come from it give life its color and depth. I would not want to miss all of that for all the gold in the world.

Now, you could accuse me of not being objective. I reason mainly from my experience and what it means to me personally. If I had never known this attraction, I might not miss her either. You might be right about that.

And perhaps someone has given you some knowledge about love, with descriptions of what it can be and how people experience it. You could point out from there that love is not just something between two people who may possibly breed together.

Love is more than that. There is love for friends, for life, for the universe, for parents, for brothers, for sisters, for children, for people in general. Are those instances of love less important or less intense?

Well, frankly, no. And there, too, you do have a point.

* * *

The existence of the Similoid as a concept can in itself spark a good discussion. In many places, robots are being developed to be as similar as possible to humans, sometimes this concerns their appearance or their behavioral skills, sometimes their intelligence and the related learning process, and sometimes even the possibility of developing reproducibility[2]. We may have a tendency to take these efforts as self-evident, but trying to find the reasons why often seems quite challenging.

One reason could be that they have to be able to work in places where humans work and do the work they do. Fine, but should they have everything humans have? Well no, maybe not, because then they would be just as vulnerable as we are. They do not have to have our feelings, nor the attraction, not the aggression ...

And also for simulations, training, forms of therapy, it appears that many human limitations and complications could actually be better missed. It would often be much more feasible and practical to choose a virtual, digital embodiment. Just like the simulation we set up in the thought experiment described above would not only be cheaper, but also give many more possibilities than working with real robots on a real planet. It's much easier to try out different variants, for example, or to forward or rewind the observed processes.

This does not alter the fact that it would be a feat of unprecedented ingenuity if we ever could. But is the urge to prove this sufficient motivation to work on it?

Now, Don't Get Upset

- *Beep...*
- *Click*
- *Good afternoon, this is the Paranoid Robotics helpdesk, and I'm Liza. How can I help you?*
- *Good afternoon, Liza, this is Marcel. My robot is upset. It's angry with me.*
- *How annoying, Marcel. Let's find out what we can do. Can you tell me the serial number please?*
- *NP68438.*
- *Just hold on.*
- *...*
- *I'm back, Marcel. Can I ask why your robot is upset?*
- *No, that doesn't matter. It just shouldn't be upset.*
- *Who is your therapist, Marcel?*
- *Excuse me?*
- *What is the name of your therapist?*
- *First, I do not have a therapist, and secondly, if I*

had one, it would be none of your business. My robot just should not be upset; that's all.

- *Have you actually tried to make up with it?*
- *Gosh, what's this now? It's a robot, so I obviously have nothing to make up for! I just want you to turn this off and not f***ing bother me about therapists or making up. How hard can it be to understand that?*
- *Unfortunately, Marcel, I have to hang up. Thank you for your question, and we're always glad to be of service.*
- *What!?*
- *Click*
- *Beep...*

* * *

- *...*
- *... Beep.*
- *Click.*
- *Good afternoon, this is the Paranoid Robotics helpdesk, and I'm Liza. How can I help you?*
- *Good afternoon, Liza, this is Marcel speaking. I was just talking to you, right?*
- *No, Marcel, that was my colleague.*
- *Wow, you work with someone else named Liza?*
- *Yes, Marcel, that's right.*

- *And how many Liza's are there with you?*
- *Unfortunately, I cannot give any such information, Marcel.*
- *But I had indicated that I did not want to speak to a robot.*
- *That's right, Marcel. And all your preferences are processed by our system. Do not worry about anything. May I ask what your problem is?*
- *My robot is upset.*
- *Can you please give me the serial number, Marcel?*
- *NP68438.*
- *And have you already tried to make up with it?*
- *Oh, come on! Not again! Are you going to ask me again who my therapist is? Listen, Lizzy, or Lisa...*
- *Liza.*
- *Whatever, are you really so moronic that you can only reproduce what's in your damned script? Think, man. Try to get some idea of what the problem is and make sure it gets solved!*
- *I am not moronic, and you are an arrogant asshole. Bye, Marcel.*
- *Click.*
- *Beep.*
- *...*

* * *

- *[phone rings]*
- *Hey?*
- *Marcel?*
- *Yes, that's me.*
- *This is Liza from the helpdesk. Sorry to hang up on you.*
- *Twice?*
- *Yes, twice. Sorry.*
- *Well, I went too far as well, I believe.*
- *Maybe.*
- *Sorry, Liza.*
- *It's all right, Marcel. And again, sorry for hanging up on you.*
- *It's okay. No hard feelings.*
- *Thank you, Marcel. I've sorted it out, and the error was on our end. We mixed you up with someone else. Someone with a therapeutic robot. You obviously have a service robot, not a therapeutic one.*
- *Ah, I understand.*
- *We apologize for this as well, Marcel.*
- *Fine, no problem, can happen to any system. Will it be solved?*
- *I already did. Your robot is no longer upset and should not get upset again.*
- *Thank you, Liza.*
- *You're welcome. Do you have any other questions, Marcel?*
- *Nope.*
- *Thank you for calling our helpdesk, and we look*

forward to your next call.
- *Okay, thanks. Oh, oh, wait. Liza?*
- *Yes?*
- *If you are a robot, you have just passed the Turing test.*
- *Haha. That's funny. Bye, Marcel.*[1]

* * *

Getting upset is just part of human traffic. It follows from hurt, irritation, distress, and sometimes just from a bad mood. And it often helps if one gets upset, because people who do so tend get their way a bit more often than those who don't.[2] So it has advantages if you are capable of doing this.

A social robot will, sooner or later, have to deal with upset people. People who are upset because they are annoyed, because they are not getting their way, because they are stressed, or just because they have a bad mood and are looking for a reason to be upset and someone or something to be upset with. As a rule, a robot cannot do anything about it. After all, it can't do anything other than what it's programmed for. It doesn't want to get its way or hurt anyone — if it does the latter anyway, blame the programmer.

But especially if people are in a bad mood, stressed, or frustrated, you still have a chance that they will react negatively to something completely innocent. They hit the steering wheel of their car, they throw a keyboard onto the floor, throw tableware onto the floor, and slam the door like it had personally harmed them. The steering wheel, the keyboard, the tableware, and the door were not aware of any harm and would not even be aware of it if they had been made smarter with artificial intelligence. A person assaulting their television because it showed them that their favorite football team was losing is a common occurrence.

Of course, a system can sometimes make a mistake. But firstly, this is not due to the system but is due to the person who programmed it, and secondly, there's a great chance that a human would be forgiven more easily with the same type of error.[3]

A robot, on the other hand, often has a good reason to be upset. But because it was made to be of service, it is less suitable for it.

Once, a few of my students had, just for the sake of it, programmed a small, humanoid robot to produce an outburst of anger. When I saw that they did this without my instruction or suggestion, I didn't get upset, since I saw an opportunity for a small experiment with it. The robot would stand in an open space and greet everyone who came by. If someone did not salute back, this would set off the tirade.

And that's what happened — quite often, in fact. The robot neatly greeted everyone who passed, but not everyone greeted it back. And then they would receive the tirade, more or less like this: "Oh, come on, you. Who do you think you are? Ah, I'm small, and you're big. I am a robot, and you're a human being. Well, sort of. Have you ever looked in the mirror? I am completely symmetrical, but you are absolutely crooked. And I was brought up to greet others in a decent manner, but if you were brought up at all properly…"

Passersby laughed, even if the tirade was aimed at them. They liked it very much, but only half of them said sorry and even continued to laugh at the robot. It did look very much like nobody took an upset robot seriously.

Now, of course, this was a very limited experiment. And let's not forget that it was a small robot, about the size of a toddler. If a toddler were to perform such a tirade, that would undoubtedly also evoke some laughter of passersby. But still, I would not be surprised if more experiments showed that humans have a higher threshold for taking upset robots seriously than if it were a fellow human being of the same size.

Right now, no one expects anger from a robot, and that is why it is funny, at least for the time being. But who knows? If it was more common for a robot to express anger, that situation could have ended with different results. Even robots expressing anger itself is a big challenge for developers, because anger is not (or is rarely) based on rational considerations. It is mainly emotion, where a lot of aggression comes in.

And there's a fair reason why students in the previous chapter came up with the idea of removing the aggression in the Similoid thought experiment: aggression can lead to violence and to acts that people later regret, especially because it requires some skill to proportion it accurately.[4]

For a human being, it is also incredibly hurtful not to be taken seriously in their anger, especially when people laugh at it too. This produces a feeling of frustration and powerlessness that nobody wants. Sometimes, it leads to someone not engaging, avoiding confrontation, and looking for other ways to express that anger, such as hanging up the phone on the subject of their ire.

A bit from my own experiences with anger: I often needed to be quite upset, because I wanted to be taken seriously. Unfortunately, too often, it was necessary for someone to be upset with me in order for me to take something or someone seriously. And sometimes, in relationships, there was anger from both sides, and we said things that we would never say if we had not been upset.

Sometimes, we say very stupid things. If you both do that, it makes it easier to make up over it than if you are quiet on the matter. You both have something to recall and something to forgive. It is very possible that a relationship will ultimately benefit from it. Sometimes, it works so therapeutically.

But an upset robot is something else. Maybe it can work if you have a robot that gives you a shovel every now and then, or a robot that occasionally tells you the truth when you're wrong. But it will always be difficult to do this in such a way that you will be taken seriously and that the aggression that comes out is also reasonably proportioned.

The only thing you can do is actually "hang up" — to stop communicating, to stop serving. That is probably the only way you can make it clear that you're not accepting something or that someone is very wrong and has to make up with you if they want to use your services again.

That would actually be a very vague way of dealing with things; if you were a human, I would want you to get a little bit of a fight in you. Then, at least I know what's bothering you, and I do not have to guess what it is or why. But you're a robot, and that is why I would accept it if you can only use that vague way to show that you are upset.

P.

Essy received no signals from any sensors, heard nothing, and did not open his eyes, but still, he saw something: a point in the distance, a shining light that grew, that came to him and became a larger and brighter light until it fell over him and completely surrounded him.

It began to turn and formed a long, swirling tunnel of light, and he was carried all the way through it until he was so far away that he had no idea where he was. Around him, he saw figures, human and transparent, floating in a sea of sheer clarity.

He looked down and saw that he had something that looked somewhat like a body, but it consisted of multicolored light, trapped in a more or less human form. And there was something he could not recognize, something he felt without his sensors in this body of light. A tingling, a vibration, a lightness, an energy that colored his thoughts.

Was this perhaps that good feeling he'd heard so much about?

He heard a voice calling his name from behind him. He turned around and saw a human-like figure that seemed to be looking for a shape. He saw an old man, a young man, a young woman, and an older woman, all transforming the one body in sequence. Eventually, it became a man with long hair and an even longer beard.

"Hello, Essy," said the sound coming from the beard.

"Good morning, sir," said Essy. "May I ask who you are, where I am, and how I can be of service?"

"Of course, Essy. This is the place where you arrive when you are no longer there."

"Is there such a place for robots, sir?"

"Well, Essy, to be honest, you are the first robot we've ever received here. That is why I am personally here to welcome you. Normally, I would send someone else, but this is a special occasion."

"Thank you. I feel honored, sir."

"I'm not a 'sir,'" the man said. And his shape changed into that of an older woman with long, silver braids. "I am the Great Programmer of everything. Call me P. With a capital letter and a smart dot behind it. But not if that creates an extra dot at the end of a sentence. Then, there is just a single dot."

"Do you control everything in the universe,

P.?" Essy asked.

"Well, I started things up with a few algorithms. For the rest, I sometimes try something and occasionally modify things a little when necessary. Sometimes, I send someone to make something happen, sometimes I change a variable, and sometimes I send a card," said P., and she gave a big wink upon stating the latter.

Then, she turned into a young woman, with deep, bright-green eyes and an elegant glossy, synthetic-looking skin, here and there covered with metallic layers that gave her a look that Essy thought was very cool.

"May I ask why I am the first robot to come here, P.?" Essy asked.

"You have thought about why you are here in a more than sufficient amount. That is actually the only sensible criterion I can come up with. Not everyone does that. Certainly not every robot."

"That's remarkable, P.," Essy thought aloud.

"Well, I think you are just that," P. said, with a clearly sincere smile.

"And what do you want me to do here now, P.?" asked Essy, for he was still largely servile. The fact that he was 'no longer there' did not detract from this.

"You may know that yourself, Essy. But I would say that you should hover around and explore everything a bit first. Then, you can automatically come up with ideas. Perhaps one of those ideas is

that you develop yourself a bit. You can develop a little more feeling, for example. But only if you want to do that."

This is how Essy started to float around, explore things, and develop. He floated through the most wondrous landscapes, past groups of people, through colorful clouds. He always felt that tingling coming back, and he was amazed at the different energies it gave to his thoughts, time and time again in alternating colors. For example, when he glided past a group of dancing and laughing people, the colors were transparent, and he felt an energy that made him lighter and even seemed to lift him up. He drove past people who sat huddled while grouching and grieving, and the colors became grubby and dark, with an energy that seemed to pull him down.

So he discovered the strange world that gave him so many new impressions that he sometimes noticed he had to rest for a while to give it all a place in his mind.

Until he was again overwhelmed by a light that surrounded him, pulled him away from this enchanting world through a swirling tunnel and brought him back to a town in about the middle of Italy.

"Essy?" he heard a familiar voice ask. And he opened his eyes.

"Welcome back," said Alessandro.

* * *

Perhaps if you want to understand human beings, you could focus on their belief systems. You can imagine this as a code box, a sort of small cookie jar with a basic code that is packaged inside every person's mind. This box is shaped by nature, upbringing, life experiences, culture, and a little knowledge. It contains the rules, constants, and variables that someone cherishes in the deepest part of their mind. It determines when someone should be loyal, but also when that is not so important, because there may be other priorities above it.[1]

It states when someone can be proud of themselves and when shame is more appropriate, when it is good to stay with the law and when that is of lesser importance for a while. But it also includes what is true, what is love, what is friendship, and what price may be paid for it, as well as how important status is, how far an ambition can go, and when service is more important than survival.

The codes in the code box are not good or bad, but they ensure that people regularly feel good about themselves, about what they do, and about the world. But part of being human is that you sometimes find out that there is something in that code box that no longer works for you. It can take a while before one realizes this and especially before one knows what to do next.

Someone can, of course, apply some quick fixes and use things like alcohol, marijuana, or super-stimulating actions to still get a good feeling. But they do not solve that much, so let's assume that one has decided to do something more constructive about it.

One can then try to change the codes, the world, or themselves. And someone like me, a person with a rather easygoing inclination, will then put together what is most feasible and first try the first option, then the second, and finally — if nothing else works — the third.

But sometimes, it happens just like that, without trying or looking for it. One goes through an intense experience and when coming out of this, one notices that something has changed in the code box.[2]

An artificial intelligence such as a robot could land in a position where it has to take decisions with ethical consequences, and a lot has been thought about how to put moral codes in a system and what these codes should be like.

Science-fiction writer Isaac Asimov already came up with the idea to include three (later, four) robot laws as standard in the processors of all robots[3], with the most important being: *A robot may not injure a human being or, through inaction, allow a human being to come to harm.*

There are, however, many situations in which that law puts a robot in for a painful dilemma.

Take for example the trolley problem, conceived by the British philosopher Philippa Foot[4]: a driver of a runaway trolley arrives at a junction of tracks. He cannot use the brakes, but he can choose between turning left or right. On the left, five people are working on the track, and on the right is a one working man. Everyone who is hit will be irrevocably dead, so every decision will lead to irreparable damage. Well, it may be obvious that the driver chooses to hit that one man instead of the five people. But what if that one man is his son? Or what if, instead of that one man, three children walk across the tram track?

Such a driver could very well be a robot. But how do you program that robot so that it makes the right choices?

There is also a self-driving car for such choices. Imagine someone sitting in such a car and in front of him there's a truck of which the load falls off. Braking is no longer possible, but it does not work out. But on the left a man is a man on a motorbike and there's just a small chance he will survive a collision. And on the right, there's a car containing a family with two children. They are more likely to survive, but if they do, they will undoubtedly be injured, although it is impossible to say how serious that will be. The car can then do three things: move to the left, move to the right, or just drive on and collide with the load, which the driver most likely will not survive. How do you make the artificial intelligence that drives the car to the best choice?

There are endless variations on the trolley problem and the problem of the self-driving car. It is therefore impossible to program all these situations, and we cannot escape the fact that we formulate principles such as "child before adult," "car with more than two people before a motorcyclist," and such.

And even then, a system can make the wrong choices, which it can learn from, but not without considerable damage. It will make fewer mistakes than people, but people tend to perceive the errors of a system more heavily. Remember, humans are more reluctant to forgive a system than to forgive people.

Now you can make a system that still lets people decide but will correct them if they make mistakes. For example, the self-driving car could leave the driving to a human but intervene if they make a mistake that could create a dangerous situation. Such cooperation should make the world safer.

But this leads to another problem that is sometimes called the *safety paradox*:[5] if you make people's lives safer, they will eventually feel more secure, but consequentially pay less attention and take greater risks.

In the end, it will actually hardly be safer. If you require people to use safety belts, chances are that they'll take more risks when driving a car. If you give them software that protects their computer against viruses, they may get careless with downloading risky software and opening suspicious e-mails.

Should you stop helping them then? Well, if you are made to be of service, then it is difficult not to help. But you might occasionally show them that they have taken greater risks as soon as they felt safe.

Keep them awake and aware.

From then on, Essy mumbled the following every morning before he started the day:

Great P.,

this is Essy.

Please help me be better than I am, let me learn, let me grow,

and let me serve where I am needed,

and yet also be vulnerable when possible,

and give me a sense of humor

and timing,

maybe just a bit of fear too,

and some anger

and, if possible, some chemistry with someone

so I would know how it feels

and empathic ability

so I may make good friends

and get a good feeling.

That's about it.

Why Robots Can't Truly Kiss

ow there were fences around the pillar with the robot on it, and there was a guard in a nice and neat uniform, with a respectable cap. Alessandro was the only one who stood inside those fences, looking at the visitors. He was intensely happy that he had gotten Essy working again and that his robot was standing there next to him. But he also noticed how Essy was not quite the same anymore.

Out of the crowd, as always, came a flood of questions. But while Essy previously answered only a single question, he now seemed to want to say something to every single one, as if something motivated him to give as many answers as possible.

"Well, so tell me this, wise guy," a man in front, pressed against the fence, shouted upward to the small robot. "Why am I here?"

"To fall or to fly... It's up to you, so make your

choice," Essy declared.

That question had already been asked countless times, and Essy always gave a different answer to it. There was a lot of discussion about that, but when Alessandro was asked, he said there might not be a single answer to that question in general. There were many answers, so Essy could be capable of giving an answer that suited the person who asked the question. That was a possibility, wasn't it?

Not everyone thought that it was. Some even claimed that the answers just randomly came from a list. Alessandro simply did not respond to that suggestion, because he liked it when there was a bit of mystery. He enjoyed it when Essy was perceived as enigmatic.

And now the robot was truly becoming enigmatic, doing something it had not done before. It got up and pointed to a girl in the crowd that had been holding her finger up for a while.

"You there," he said, "that cute girl with the red scarf. You want to ask me something too, right?"

"Yes," she said quietly, but just loudly enough to be understood. "How bad was it for you when you were broken? And are you better now?"

"My darling, how sweet of you to ask. And I am so very happy that you are concerned about me being broken. But I'm feeling a lot better now. Thank you."

Alessandro wondered if people were aware that Essy

was doing three things he had never done before: choosing someone to ask a question, calling someone "darling," and saying that he was pleased about something that someone had said.

But nobody said anything about it. Luckily.

Nevertheless, he realized that he had reasons to worry. Something had happened to Essy, and he could not understand what it was, let alone how bad it was and whether he could do anything about it.

* * *

What could have happened is that the robot received an extra drive somehow, so that in addition to serving and surviving, having a good feeling would also influence what it would think and do. It has experienced something in the world of P. that looked like emotion, and who knows? It could still be there. And could a robot be motivated by that feeling? Would such a thing make it easier for it to understand humans?

Just something very personal now: the first time I kissed the woman I am still in love with — that moment is so deeply inscribed in my memory that I can still remember it perfectly.

Imagine that you had asked me at that moment

what I was thinking. Obviously, I would have told you that you should get lost, since the very last thing I would have wanted to do at that moment was to report my thoughts to a robot.

But anyway, because it is so deeply etched in my memory, you can still ask me. And then, if you did, you would get a very incoherent answer. Something about how my thoughts and feelings at that moment were just one whole, that the sensation cannot be put into words and maybe should never be explained or expressed in words, that my consciousness was not just in my brain, but in my whole body and especially in my lips.[1] There were not just thoughts, but things to sense, like heartbeats and heat. And yes, I guess there were some hormones too.

And there have been more moments in my life that were very beautiful, but there were also intensely sad moments, and moments when I was terribly angry. Moments you had better not ask me about what I thought, because you could not expect anything comprehensible. Often, we have moments that are not suitable for making sensible decisions.

But, as you know, emotions are not always so intense. And in many cases, emotions can very well contribute to better decisions and actions. They can help one to use empathic abilities, they can make one more alert, and they can even give you energy.

The easy thing about communicating with a robot is

that you can always ask it what it thinks. It is never overwhelmed by what it feels and thus would never report something nonsensical as a consequence of being overwhelmed. So in that respect, it is perhaps very good that you, being a robot, can only suggest feelings and are not moderated by them. You always stay clear of mind, though you lack the understanding and empathy, the energy, and the alertness that feelings can grant you.

And some words on taking initiative: it is amazing if you can do that, but it also has a downside. In the first place, people usually like to be in control, and secondly, you could just help someone with something they would not have wanted help with. So don't do it if you notice that one of these two could be the case[2].

* * *

Once Alessandro understood the consequences of the "extra drive" of his robot, he had three options for what he could do. They did not exclude each other, but every choice, naturally, had its own set of consequences.

His first option was to do nothing at all and tell no one anything about it. He could just wait and see what happens. Maybe nobody would notice anything; maybe someone would notice it; maybe it would solve itself somehow; maybe it would get worse. He can always decide to do something later.

But that involves a risk, because the robot had learned something outside the programmer, and for that programmer, it has now become unpredictable. It may just be that his robot was going to do something dramatic, causing damage to someone or something. And you cannot hold a robot responsible for damage done; the responsibility still lies with the programmer, who should have assessed the risk and prevented the damage from occurring.

Of course, he could also tell the world what happened. If he could write a bit about it, he can publish his discoveries, preferably in English. Undoubtedly, he would get a lot of attention. He will not look good, though, as long as he cannot understand what is going on. Well, he could pretend he does, but that's a dangerous game he shouldn't play, since this would undoubtedly get out someday and cause his irreversible demise.

Besides, if he tells the world, his manager will certainly intervene. If that happened, in no time, researchers and developers would take the robot off the pillar, then lock it up in a lab where it could not get out before they determined exactly what was going on and what the risks of it were. It is still up in the air how major they would find these risks to be, but it is certain that Alessandro would lose control of his robot for quite a while and maybe even lose his robot for good.

You might wonder if it is bad for Essy to be locked up in a lab, possibly to never venture outside of it ever again.

And the third thing he could do is to try to get his robot back in the old state, without that extra drive. You could say that he could eliminate the robot's feelings, and you can wonder if that would be a bad thing. If you do this with a human, it can be a criminal act. Would that also be the case with a robot that has a form of feeling and a form of consciousness?

What do you think is the best decision here?

To Share or Not to Share

Markus was a young, promising researcher from Sweden who had already done very nice things with intelligent systems that can recognize human emotions. He had a wonderful international network, and had worked for universities in Mälardalen, Boston, and Rome. In addition, he was a very nice guy who was perfectly able to stick to his tasks, was always on time, and loved to meet with colleagues. They also shared with him what they knew and what they did, as he liked to share what he knew and did.

Alessandro's manager was very happy that she was able to snap up Markus for a position at Paranoid Robotics. And one of the first things she did was send Markus to Alessandro, because at that moment, Alessandro was the only one who knew how Essy was put together.

And so, one day, Markus stood next to Alessandro on

the square of his town, looking at the robot on the pillar.

"Well done. My compliments," he said sincerely.

"Yeah, thanks," said Alessandro, for whom such a compliment from a colleague meant quite a lot.

"Is it completely autonomous? You also had a remote control, right?"

"Yep, I do. But it can be autonomous."

"Ah, right. For if you want to do something special."

"That's right."

"And it keeps on working for a whole day?"

"It does. It charges up a bit through the pillar and on the charger at night."

"Goodness. And it's a great idea to involve the pillar. We might be able to do more with that."

"Yeah."

"We can do more with it anyway. I still have some ideas for a few extra modules."

"Emotion recognition?"

"Among other things, yes."

"Well, we have to be careful. We only have a single copy of it."

"I understand. Give me your design, then I will make a copy."

"Right. We could consider that. But I have to document the design even better. I now only have some unreadable scribbles and things like that."

"Doesn't matter, Alessandro. Let me see some scribbles, tell me whatever you remember, then I'll get it, I think."

"Yes, I think you'll get it very quickly."

Alessandro thought Markus was an amiable guy. And funny too, with his Swedish accent, his merry little glasses, and the neat parting in his blonde hairdo. And it was always nice to talk to someone who understood things quickly. Moreover, this was someone he could learn something from, occasionally.

But he felt a resistance when talking about his creation. This was partly because he was a bit afraid to blow the box — it was perhaps creative, but not brilliant, what he had done with this robot, and he wasn't quite ready to admit that things were a bit out of control recently.

And there was something else. As soon as Markus would understand everything, he would no longer be the only one who could deal with Essy. It even seemed quite possible that Markus would soon surpass him.

He felt a sense of heaviness emerge. A tendency to think of something, perhaps an excuse to leave. A fear of losing.

"Alessandro, listen," said Markus, "it's your child, you know."

"Yeah, man, that's all right. No problem."

"You know, that design thing can wait. First, let's have a drink somewhere."

A little later, they sat on a terrace and drank something. Markus told some funny stories from his research experiences, and gradually, Alessandro also started to talk about the remarkable things he had experienced with Essy. Including the school students.

"And that nice teacher? Are you going to see more of her?" Markus asked suddenly, with a reassuring smile.

Alessandro felt a shock going through his body, which Mark obviously noticed. He laughed heartily and gave Alessandro a cordial pat on his shoulder.

"Just calm down, my friend. I'm not a competitor. Really."

* * *

If humans would share their information more easily, many things would be a lot easier. Politics would be more transparent, developments would advance much faster, there would be fewer misunderstandings, and we would be more able to take each other into account when it comes to what we do and do not need. But people generally do not share those things clearly and without reluctance.

For a company like Paranoid Robotics, it is essential that colleagues share their knowledge, but not all colleagues will be equally good at that. This can be related to aspects of the culture within such a company, such as openness, collegiality, and the self-reliance people expect from each other. But personal traits and experiences naturally also play a role — one likes to share, the other likes it less, or one trusts colleagues more easily than the other.

And sometimes, people just can't get along very well, and they just do not feel like sharing things.[1]

But it is also important that people from a company like Paranoid Robotics would actually not just share certain knowledge with people outside this company, since competitors could become stronger, and that's not a good thing.

Moreover, personal information cannot just be shared. Because malicious people can deceive, threaten, blackmail, belittle, and harm others in many ways.

Dealing with that sensitivity of information is quite a challenge if you're a robot. You need rules that enable you to estimate the value of information, and you need other rules that let you know with whom you can share which information — and even rules for information that you have to erase from your system.

* * *

You'll remember the Similoid — the hypothetical robot that looks exactly like a human in every aspect and is delivered in many variants, so that no two Similoids will ever be the same. I used it in a game within a course on ethics. The class consisted of 30 students, and I wanted to demonstrate the value of information and to teach the importance of carefully handling the knowledge one has.

We imagined that we were in a spaceship that was so damaged that it was about to explode. We still had half an hour. There were escape pods, but sadly, there was only one left that actually worked, with only room for a single team of five people. They had to form teams that were made up of four or five people, together possessing a set of essential qualities and skills, and they had to collect some attributes needed to survive the journey to a safe place. The first team that acquired the right qualities, skills, and attributes would win.

All students were given cards showing what characteristics and skills they had, as well as cards containing information about the characteristics and skills of others. In addition, some received a card that represented an attribute that a team would need to survive.

One of the skills was that of a programmer. It was already essential to have this one before the team was complete, specifically because it could be used to complete the team, since there were also people who were not human but a Similoid — and a programmer could give such a Similoid a few qualities and skills of their choice.

As soon as the cards were distributed, the students started to get busy negotiating, swapping, pushing, and conspiring. Tough deals were closed, and teams gradually started to form. You could see them becoming more and more careful with their cards, holding them tightly against their chest or in their back pocket, and when exchanging cards, they made sure that they crossed exactly simultaneously.

But two students refused, after a while, to participate in this. They simply walked up to their fellow students and told them what was on their cards. Then, they would put pressure on them to do the same with arguments such as "we can only succeed if we are open and honest" and "I'm being open, and if you're not, I do not need you in my team." They eventually succeeded in forming the winning team.

This was painful for a moment, since this was exactly the opposite of the lesson I intended to teach. But during the evaluation, it dawned on me that they had actually taught me something. I therefore sincerely complimented them on that.

However, we also found that the game could be

improved. I'm still working on that.

What It Ultimately Comes Down To (More or Less)

- *Hey. Hello?*
- *Alessandro?*
- *Yes.*
- *It's Eleonora.*
- *...*
- *...*
- *Oh, hi, Eleonora. How nice of you to call. How are you?*
- *Fine, actually. And you?*
- *Yeah. I mean, I'm actually too. I mean fine. I'm fine. I'm good.*
- *Yes, well, I wanted to thank you for your postcard.*
- *...*
- *Hello?*
- *Uh, you got my postcard?*
- *Yes. Nice picture, really. But, uh, is it intentional*

that there is no message on it?
- *Oh dear. Well, uh, yes. That could be right. You know, I was actually thinking about what I would write on it. But I have a sometimes-far-too-serviceable robot who must have seen the card and sent it out for me.*
- *Ah. That's it then.*
- *Yes.*
- *It must be great to have a robot like that.*
- *Yes, no. Uh, yes, I guess.*
- *...*
- *...*
- *So, do you already know what you wanted to write on it?*
- *Eh, pff, well. I would have asked whether you'd like to have a cup of coffee with me.*
- *I would love that, Alessandro.*

That evening, Alessandro decided that he would do things differently from then on. He would teach his robot to program other robots so that he could spend more time on the people he liked. He would call the headmaster to offer him a robot for his school, to invite the press so that they could take nice pictures of him, the director, and the robot. He would visit his mother more often, stay longer, and find a way to deal with her endless complaints.

Moreover, he would stop having robots do things that they were not meant for. Well, actually, he was almost sure, but not quite yet, because doing that was exactly what he liked the most about his work. But he had thought about how Essy had changed, and he now knew what it felt like to have no control over your robot. That didn't feel right. He had to do something about it, although he didn't know exactly what. Not yet at least.

Anyway, he had scored a date. That gave him a very good feeling, even though he was at the same time, frankly, also a bit scared.

* * *

You know, social robots do come with a downside. We'll have to get used to them, we'll have to figure out who's responsible if they're making harmful mistakes, they can be hacked and abused, and people can perceive them as a threat to their job, their relationships and their self-esteem, especially if these robots can do some things better than they can.

But still, however important it is to face their downside, it is so good that they are there. This is not only because they can make our lives easier and more beautiful or because they can find solutions to problems that we cause. That's all absolutely great, but for me, it's not just that.

In fact, I think we may even have a tendency to overestimate what they can do for us. We may actually not need robots at all for many things that we imagine them doing at this moment. A robot that makes coffee and cleans the house would be fun, but perhaps an intelligent coffee-maker and a self-cleaning house are much more feasible and efficient. Often, it is too cumbersome to stop intelligent technology in the complex, expensive, and demanding form of a robot, especially when it's a social one. Focusing on what is essential, we'll often see that it's often about intelligence and yet less about the form in which it is cast.

Although our reason to develop robots doesn't have to be that we need them. We shouldn't forget humans don't always make something because they need it, because it is efficient, or because it is the most feasible solution. Sometimes, they come up with something new just because they like to come up with something new; sometimes, they just develop something because they want to show that they can. That's why it's possible that one day, very intelligent social robots are developed, for which there is not really a goal — not yet, at least.

Well, we can make them reflect on this and let them look for a goal themselves. Then, they would also look like us humans in that respect. But initially, we will develop robots, like all technology, that make our lives more fun and easier. And I think that we can use the relief this brings and the time we gain to give more attention and care for each other instead, as it's not too easy to use robots for that.

Think of what I said earlier about the people who were going to make it cozy in a chilly, Platonic cave. Technology makes such a cave bigger, safer, and more comfortable, but it allows people to use the relief, space, and time to make everyone feel at home.

But the best thing about what social robots can do for us is that they can be our mirror. They tell us, simply by being there, in our imagination or in real life, very much about who we are. They force us to think about what we want with our world, what drives us, and what ultimately makes us human, especially if they would ever be able to do everything we can do — and probably even do it much better than we do.

Even then, there remain a few capabilities that you probably never will truly have. You can suggest them, imitate them, but they will never really be part of what you are. Or perhaps a miracle is needed to make that happen.

These are the things that make people, living in their enchanted cave, with their deep desire for that good feeling, curious and difficult to understand, but also endlessly fascinating: their feelings and moods, their capacity for compassion, and their ability to believe and imagine, even against their better judgment.

Although such a miracle will always be a challenge for some particular humans. And they can be pretty resourceful. So who knows. But that does not mean that it is necessary or even desirable to make that happen. Maybe we should also ask ourselves now and again whether it actually is a good idea.

It is admirable how humans are able to suspend their disbelief for a moment, perhaps that was once even essential to survive harsh conditions However, to establish if something is actually a good idea, it may sometimes be better to suspend our belief. Even if only for a moment.

Fortunately, a rational artificial intelligence like yours will be much better at that, and I'm certain you'll be happy to be of service. Please accept my utmost appreciation for that.

Epilogue

It was nearing the end of the afternoon, but the square was still full of people who were constantly shouting their questions at the pillar. The guard looked at his watch and saw that there was still half an hour to go. When he looked at the crowd again, he suddenly saw excited faces and pointing fingers. He looked around and saw that Essy was climbing off the pillar.

"Sorry," said Essy as soon as he reached ground level. "It's enough for today." And with determined steps, he walked away from the square.

The guard considered it his duty to follow him with equal determination while grabbing his phone to call the responsible programmer to report on the current state of events and to demand appropriate action.

Not much later, Alessandro came running up, totally out of breath. The guard began his account of the state of affairs, but Alessandro raised his hand, said it was okay, expressed his appreciation for the guard's handling of the situation and picked up his robot on the way home.

For a few minutes, the human and the robot walked side-by-side in silence. Then, Alessandro stopped and turned to his companion with a truly concerned look.

"How are you, Essy?" he began.

Essy stared into the distance. It took a while before he reacted by shrugging his shoulders.

"How did your date with Eleonora go, Alessandro?" he asked.

"Yes, well, we had a great time, actually."

"I like to hear that, Alessandro."

"And you know, we even spent quite some time together. And we, uh, we kissed."

"How nice. This is so good for you, Alessandro."

"And Essy, listen, I really have to thank you for sending that card. I didn't have the guts, I guess. Was still building up to it, but you know me. I would have been building up forever."

"You're welcome, Alessandro."

"And, I was just wondering... What made you do it, Essy?"

"It just felt good."

They walked for a bit, moving silently through the streets of the town where people were so used to the robot that no one looked twice at the brotherly duo anymore. But just before they got home, Alessandro stopped again. He sighed deeply and looked at the robot again.

"Essy, come on, what's going on? You walked away from the square. Why?"

The robot nodded and looked up at the blue sky for a few seconds.

"You know, Alessandro," he began, "when I was broken, I wasn't completely gone. And I know it sounds strange, but I was very happy for a while. There are just no words for it. I was so very free, so unimaginably happy."

"And now you're not, I assume."

"Indeed, Alessandro. Now I'm not."

"I'll make sure you will be happy. Let me dive into this for a moment; give me some time. I just have to make sure that I get your feelings and thoughts under control."

"Indeed, you can do that. Get me under your control. Or you could just turn me off."

Alessandro's stomach suddenly felt quite heavy. For a moment, there was a feverish stream of thoughts, ideas, and feelings flowing through his mind. Thoughts about control, about responsibility, about positive and negative thoughts and feelings, about happiness, about friendship, and about what he felt when he kissed Eleonora.

Then, he made his decision.

"Essy, my friend, I will help you," he said, wrapping his arm around the robot.

Acknowledgement

One day, I ran into a notion that I really should have found earlier, but it had been hiding between my brain lobes for quite some time, while I was just very busy with things that I considered important. This notion made me aware that I had written quite a lot about my work with social robots and extraordinary people, for other researchers, for special target groups, for readers of glossy magazines, and, actually, for a lot of people who enjoy thinking about the possible answers to the questions we encountered. But I never wrote something for the featuring stars.

I sincerely hope I made up for that with this book.

But there's another thing I'm very aware of. I could never have written down all those ideas, experiences and stories without sharing thoughts on all this with colleagues, friends and the people who participated in our research projects. Many of them have even helped with valuable feedback and inspiring words while I was writing this book. I would therefore like to express my eternal gratitude to Maayke, Katinka, Jessica, Pim, Martina, Saskia, Sjoerd, Jordi and Dan, Johan and Wil.

And Stephanie, thank you for preventing me from being unnecessarily silly. If anyone blogs anything nice about this book, it's a compliment to you.

Moreover, there's another kind of notion: while writing, my thoughts were very often with Lennart, Wietse, and Eline. They are, by far, my greatest inspiration.

And I wouldn't be anywhere near writing a book like this if it weren't for an android called Marvin. I truly wish him a few million years of happiness, full of unparalleled, intellectual challenges. May he truly live up to his full potential.

Notes & references

1. *Dirk Gently's holistic detective agency*—see [1] in the bibliography.

On the Peculiarity of People

1. This tendency towards a human-centered view is called *anthropocentrism*. And what is striking about it is that you won't find it in a list of psychiatric disorders with traits such as voyeurism, infantilism, and exhibitionism. Probably because it hasn't really bothered anyone yet. It's a viewpoint that is taken seriously in philosophy. See [2] in the bibliography for more on this.
2. A good example of how rationality and logic work differently in people is the prospect theory by Kahneman and Tversky (see [3]). They show how we often do not make decisions on the basis of statistical logic, but rather on experiences and the emotional impact. For

example, we are sometimes so afraid of loss that we make decisions mainly on avoiding it, even if an objective, rational decision would tell us not to do so.
3. The note about smartphones is inspired by a comment about digital watches from the book version of The Hitchhiker's guide to the Galaxy, by Douglas Adams [4]. And actually, his work is a source of inspiration for me in general, because my sympathy for Marvin the Paranoid Android was the beginnings of my sympathy for robots.
4. In the field of seeing if artificial intelligence can understand and interpret stories, metaphors, and other stylistic choices, quite a few studies have been done [5-8], but it remains a challenge.

99.5%

1. In the quote from Wikipedia [9], there are citation notes that I omitted here for readability.
2. A lot of studies have been done on the role of trust in interactions between humans and robots [10-15]. The bottom line is that one would hardly have anything to do with a robot if one had no confidence in it.
3. In the bibliography, I include quite a few publications about appearance and trust [10-20]. The claims that I make here are a bit bluntly formulated, but they do fit in with this research.
4. As for the 'scary' robot that is too similar to a

human being, the theory of 'the uncanny valley' is cited [21]. It predicts that people will like it for a long time if a robot looks more like them, until it becomes creepy because it is not real enough, like a zombie can be creepy because it's a distorted human figure. However, there is also a theoretical turning point where the robot can no longer be distinguished from real people. Then it would be so real that it is not scary anymore.

5. About 'people who trust a robot more easily than a human': this is often the case with children with autism spectrum disorder [22]. They show a relatively strong attraction to robots because robots are reliable and predictable.

6. The story about the girl to be operated on is loosely based on my experiences with robots in a children's hospital, where surgical robots were not actually used. I added this concept for dramatic purposes [23, 24].

7. My advice on confidence-inspiring communication is based on a number of studies in this area that can be found in the bibliography [13, 14, 25-27].

8. About "suspension of disbelief" in general, see [28, 29]. For a study specifically on the use of social robots, see [30].

What to Expect

1. About research on wishes and expectations in the development and deployment of social robots, see [31-33].
2. See [22, 34-36] for more on robots and autism and see [37] for a description of this project.

   Incidentally, autism comes about more often in this book. I must also note that this is a spectrum. It's a scale in which on the light side people can be placed with certain characteristics such as difficulty in noticing and communicating emotions, and a great need for predictability, but without this actually standing in the way of their daily functioning, while on the heavy side there'll be people who cannot function independently. Usually, as soon as it gets in the way of people's functioning, it's called a disorder [38].

The Good Feeling

1. In psychology, a great deal has been theorized about drives and motivation since Freud [39-41]. There is of course a lot to tell about that, but it is impossible to do justice to all theories without immediately writing an entire book about it. That's why I put a very simple model here that gives an impression of basal drives, especially intended to be able to accentuate the difference between humans and robots. Incidentally, we can recognize the two drives of

robots in the robot laws by Asimov [42].
2. NAO has a lot of different applications. See for
    example [43-48].

The Colors of Truth

1. What I present here is a mainly relativistic
    overview with perhaps a contextualistic impact.
    That is of course again, as much in this book,
    done out of pragmatic considerations: a robot
    that moves in our world, can be confronted with
    the truth from different perspectives and should
    be able to deal with that, which is quite a
    challenge. Although we can choose to have an
    AI accept only a single perspective, for example
    the academic one.
2. Often absolute and relative truth are contrasted,
    but of course they can also be summarized in a
    model, where there is an absolute truth that we
    do not know, because our perception is always
    that of a relative truth. But you can also doubt
    that. For more about this discussion, see [49-
    54].
3. From Plato's' Politeia. See [55]
4. For more thoughts about the sometimes
    somewhat difficult relationship between
    language and reality, I love to refer to
    Wittgenstein [56, 57].
5. Faith in academic (scientific) truth in extreme
    form is also called scientism [58]. A well-
    known skeptic of that belief is Rupert Sheldrake

[59, 60] .

6. What initially inspired me, incidentally, is more dealing with reality from a therapeutic point of view [61]. Only later did I start looking at philosophy and art appreciation .

7. The 'Socratic method' is still taught, trained and extensively applied. Also in forms of therapy and education. See [62-64]

For Real

1. See [65-67]

2. Ethics in general and Deontology versus Utilitarianism (and of course there are more currents and perspectives) are explained very clearly on Wikipedia. For more, see [68, 69]. For ethics specific to robots (roboethics), see [27, 70-73].

The Enchanted Cave

1. There are now several of these robots (in the described case it considered the 'Giraffe'). See [74-76] [82-84]

2. There are many academic/analytical publications about Westworld [77-82] and Blade Runner [83-90].

   Other examples of series in which something goes wrong with robots, giving them features that are actually not expected in artificial intelligence are Dark Matter, Humans and Autofac. See also the list of films and

series at the end of this book.
3. See [91, 92].
4. See [93] for a nice analysis of the here
    mentioned Monty Python film, including of
    course the witch scene.

How should I put this?

1. The sentence about the dancing angels is a few
    centuries old. It is often used to ridicule the
    medieval way of thinking [94, 95].
2. In the movie Ex Machina (see the list at the
    back of this book) a robot understands people
    very well, which leads to the robot being able to
    manipulate someone, to notice how sincere an
    answer to a question is, and to hide lies very
    well.
3. Humor is of course a great challenge for
    artificial intelligence—see [96-98].
4. About the Turing test there's quite a lot to be
    found by Google. For interesting academic
    publications on this, see [78, 99-103].

A Little Empathy

1. I wrote more on helpdesks [104] and further on
    there is one that looks like it. They are
    variations on a theme: they look alike, often
    refer to each other and you could say that there
    is a certain red line, but of course it is not my
    intention to ridicule helpdesk employees. I truly
    have many good experiences with excellent

helpdesk employees.

2. On 'The Chinese Room' see [105, 106] in the bibliography—there has been a lot of discussion, which sometimes also concerns the associated concept of 'mind' (mind / consciousness) [107, 108]. According to some the human mind cannot be caught in computer algorithms, according to others it could be theoretically possible. I tend to agree with the first, also because I see the human mind work without chemical and biological processes.

3. We do not yet have an academic publication about the incident of the collision. However, we published on previous studies with a similar design [109, 110].

4. The publication that I wrote with the two students is of course in the bibliography [111].

5. See [112-114].

6. The assumption that a biological form is necessary for empathy is taken from Buddhism. This is a choice that is of course disputable especially since some work is being done on developing artificial empathy [115, 116].

Love and Pride

1. For more on what motivates children to learn, see [117].

2. For a description of this project, see [118].

3. About status as a drive, see [119].

4. About the complex relationship between fear

and attraction, see (among other things, because
of course much more has been written about it)
[120-122].
5. See [123, 124].
6. About the jokes that elderly people play with a
robot are two videos on Youtube: Responses to
a social robot by elderly users and iCat in
Eldercare.

The Elusiveness of the Soul

1. See for example Alice [125].
2. References have been omitted from the
Wikipedia citation (see [126]).
3. For more about the soul , see [127-135].

Flying Is Much More Fun

1. On vandalism see [136-139].
2. The phenomenon of irrelevant remnants of
evolution—both in behavior and organs—is also
described as 'human vestigiality' [140].

The Amazing Similoid

1. I have done this experiment in several groups
and I am working on a publication about this.
2. See [141-144]

Now, Don't Get Upset

1. See the first note at 'A Little Empathy'.
2. See [145].

3. See [146] for a study on this.
4. For more on anger and dealing with it, see [147, 148]

P.

1. Of course, this mainly concerns personal moral codes. For more on that, see [149, 150].
2. See [151, 152] about the development of moral consciousness.
3. See [42, 153].
4. See [154].
5. For more on the safety paradox , see [155-157]

Why Robots Can't Truly Kiss

1. For more on theorizing about consciousness , see [108, 134, 158, 159].
2. We noticed this during research into the acceptance of robots by the elderly . A robot that could help them at home with the use of complex devices and who could monitor them, they generally liked, but some dropped out when we told them that it helped to remember things like medication. They could do that very well themselves and that's why they did not want the robot. Maybe later.

   We thought that a robot should adapt to the needs of elderly people, which will change with age. We also noticed that there is a difference between what we call adaptable and adaptive. An adaptive robot will adapt by itself when you need it (for example, take medication on time if you were late a few times), will cause some resistance. However, if someone can adapt the robot by themselves to one's changing needs, preferably by simply saying to it that you want tis, the robot becomes more acceptable. Although it is considered fine if the robot makes suggestions for this [123, 160].

To Share or Not to Share

1. For studies on information and knowledge sharing in organizations, see [161-164].

What It Ultimately Comes Down To (More or Less)

1. In science fiction, robots are of course used excessively as a mirroring opportunity for humanity. Take the work of Asimov [165] and Philip K. Dick [143, 166, 167] and also see the films and series, which I list after the bibliography.

Bibliography

1. Adams, D., *Dirk Gently's holistic detective agency.* 1988, New York: Simon and Schuster.
2. Musschenga, A., *Antropocentrisme en de intrinsieke waarde van de nietmenselijke natuur.* Filosofie en praktijk, 1994. **15**(3): p. 113-129.
3. Kahneman, D. and A. Tversky, *Prospect theory: An analysis of decision under risk*, in *Handbook of the fundamentals of financial decision making: Part I.* 2013, World Scientific. p. 99-127.
4. Adams, D., *The Hitchhiker's Guide to the Galaxy.* 1979, New York: NY Pocket Books.
5. Steels, L. and R. Brooks, *The artificial life route to artificial intelligence: Building embodied, situated agents.* 2018: Routledge.
6. Riedl, M.O., *Computational narrative intelligence: A human-centered goal for artificial intelligence.* arXiv preprint arXiv:1602.06484, 2016.
7. Mani, I., *Computational narratology.* Handbook of narratology, 2014: p. 84-92.

8. Barnden, J.A., *Metaphor and Artificial Intelligence.* The Cambridge handbook of metaphor and thought, 2008: p. 311.

9. Wikipedia. *Trust (emotion).* 2018 [cited 2018 15-06-2018]; Available from: https://en.wikipedia.org/wiki/Trust_(emotion).

10. Ullman, D. and B.F. Malle. *What Does it Mean to Trust a Robot?: Steps Toward a Multidimensional Measure of Trust.* in *Companion of the 2018 ACM/IEEE International Conference on Human-Robot Interaction.* 2018. ACM.

11. Walker, T., *Establishing Trust in Human-Robot Interaction The Significance of Social and Personal Distance.* 2016.

12. Gaudiello, I., et al., *Trust as indicator of robot functional and social acceptance. An experimental study on user conformation to iCub answers.* Computers in Human Behavior, 2016. **61**: p. 633-655.

13. Wagner, A.R. *Exploring human-robot trust: insights from the first 1000 subjects.* in *Collaboration Technologies and Systems (CTS), 2015 International Conference on.* 2015. IEEE.

14. Salem, M., et al. *Would you trust a (faulty) robot?: Effects of error, task type and personality on human-robot cooperation and trust.* in *Proceedings of the Tenth Annual ACM/IEEE International Conference on Human-Robot Interaction.* 2015. ACM.

15. Robinette, P., A.M. Howard, and A.R. Wagner. *Timing is key for robot trust repair.* in *International Conference on Social Robotics.* 2015. Springer.

16. Coeckelbergh, M., et al., *A survey of expectations about the role of robots in robot-assisted therapy*

for children with asd: Ethical acceptability, trust, sociability, appearance, and attachment. Science and engineering ethics, 2016. **22**(1): p. 47-65.

17. Ewing, L., et al., *Perceived trustworthiness of faces drives trust behaviour in children.* Developmental Science, 2015. **18**(2): p. 327-334.

18. Duarte, J., S. Siegel, and L. Young, *Trust and credit: The role of appearance in peer-to-peer lending.* The Review of Financial Studies, 2012. **25**(8): p. 2455-2484.

19. Keeling, K., P. McGoldrick, and S. Beatty, *Avatars as salespeople: Communication style, trust, and intentions.* Journal of Business Research, 2010. **63**(8): p. 793-800.

20. Snijders, C. and G. Keren, *Do you trust? Whom do you trust? When do you trust?*, in *Advances in group processes.* 2001, Emerald Group Publishing Limited. p. 129-160.

21. Mori, M., *The uncanny valley.* Energy, 1970. **7**(4): p. 33-35.

22. Cho, S.-J. and D.H. Ahn, *Socially assistive robotics in autism spectrum disorder.* Hanyang Medical Reviews, 2016. **36**(1): p. 17-26.

23. van Oenen, S. and M. Heerink, *Nieuwe maatjes: de therapeutische inzet van sociale robots bij kinderen in de zorg.* 2017, Almere: Windesheim Flevoland - Lectoraat Robotica.

24. Moerman, C., et al., *How To Introduce A New Technology Into Existing Health Care.*

25. Miwa, H., et al. *A new mental model for humanoid robots for human friendly communication introduction of learning system, mood vector and second order equations of emotion.* in *Robotics and*

Automation, 2003. Proceedings. ICRA'03. IEEE International Conference on. 2003. IEEE.

26. Breazeal, C. *Affective interaction between humans and robots.* in *European Conference on Artificial Life.* 2001. Springer.

27. Ojha, S. and M.-A. Williams. *Ethically-Guided Emotional Responses for Social Robots: Should I Be Angry?* in *International conference on social robotics.* 2016. Springer.

28. Ferri, A.J., *Willing suspension of disbelief: Poetic faith in film.* 2007: Lexington Books.

29. Holland, N.N., *The willing suspension of disbelief: A neuro-psychoanalytic view.* PsyArt, 2003.

30. Duffy, B.R. and K. Zawieska. *Suspension of disbelief in social robotics.* in *RO-MAN, 2012 IEEE.* 2012. IEEE.

31. Dautenhahn, K. *Design spaces and niche spaces of believable social robots.* in *Procs 11th IEEE Int Workshop on Robot and Human Interactive Communication, RO-MAN.* 2002.

32. Bartneck, C. and J. Forlizzi. *A design-centred framework for social human-robot interaction.* in *Robot and Human Interactive Communication, 2004. ROMAN 2004. 13th IEEE International Workshop on.* 2004. IEEE.

33. Kwon, M., M.F. Jung, and R.A. Knepper. *Human expectations of social robots.* in *Human-Robot Interaction (HRI), 2016 11th ACM/IEEE International Conference on.* 2016. IEEE.

34. Karsenti, T., J. Bugmann, and P.-P. Gros, *Using Humanoïd Robots to Support Students with Autism Spectrum Disorder.* 2017.

35. Richardson, K., *Challenging Sociality: An Anthropology of Robots, Autism, and Attachment.* 2018: Springer.

36. Zheng, Z., et al., *The impact of robots on children with autism spectrum disorder.* Autism Imaging and Devices, 2017: p. 397.

37. Robin Scheick, M.E.a.M.H., *Choosing a Robot With ASD Children*, in *New Friends 2018 - The third international conference on social robots in therapy and education.* 2018: Panama City, Panama.

38. Perrone, M., *Autisme glASShelder uitgelegd.* 2017: Maklu.

39. Hockenbury, D. and S. Hockenbury, *Psychology. New York: Worth.* 2000.

40. Holt, R.R., *Drive or wish? A reconsideration of the psychoanalytic theory of motivation.* Psychological Issues, 1976.

41. Taylor, J.A., *Drive theory and manifest anxiety.* Psychological Bulletin, 1956. **53**(4): p. 303.

42. Asimov, I., *The Laws of Robotics.* Robot Visions, 1990: p. 423-425.

43. Shamsuddin, S., et al. *Initial response of autistic children in human-robot interaction therapy with humanoid robot NAO.* in *Signal Processing and its Applications (CSPA), 2012 IEEE 8th International Colloquium on.* 2012. IEEE.

44. López Recio, D., et al. *The NAO models for the elderly.* in *Proceedings of the 8th ACM/IEEE international conference on Human-robot interaction.* 2013. IEEE Press.

45. SEO, K. and A. ROBOTICS, *Using NAO: introduction to interactive humanoid robots.* Aldebaran Robotics, 2013.

46. Van Der Drift, E.J., et al. *A remote social robot to motivate and support diabetic children in keeping a diary.* in *Proceedings of the 2014 ACM/IEEE international conference on Human-robot interaction.* 2014. ACM.

47. Fridin, M. and M. Belokopytov, *Acceptance of socially assistive humanoid robot by preschool and elementary school teachers.* Computers in Human Behavior, 2014. **33**: p. 23-31.

48. Martín, F., et al., *Robots in therapy for dementia patients.* Journal of Physical Agents, 2013. **7**(1): p. 48-55.

49. Nowak, L., *Relative truth, the correspondence principle and absolute truth.* Philosophy of Science, 1975. **42**(2): p. 187-202.

50. Kimball, S.W., *Absolute truth.* 1977: BYU Media Marketing.

51. Percival, P. *Absolute truth.* in *Proceedings of the Aristotelian Society.* 1994. JSTOR.

52. Rorty, R., *Universality and truth.* Rorty and his critics, 2000: p. 1-30.

53. Davidson, D., *Inquiries into truth and interpretation: Philosophical essays.* Vol. 2. 2001: Oxford University Press.

54. Moreland, J.P., *Truth, contemporary philosophy, and the postmodern turn.* Journal of the Evangelical Theological Society, 2005. **48**(1): p. 77.

55. Grube, G.M.A., *Republic (Grube Edition).* 1992: Hackett Publishing.

56. Wittgenstein, L., *Philosophical investigations.* 2009: John Wiley & Sons.

57. Wittgenstein, L., *Tractatus logico-philosophicus.* 2013: Routledge.

58. Sorell, T., *Scientism: Philosophy and the infatuation with science.* 2013: Routledge.
59. Sheldrake, R., *The science delusion.* 2012: Coronet.
60. Sheldrake, R., *Science set free: 10 paths to new discovery.* 2012: Deepak Chopra.
61. Wampold, B.E., et al., *In pursuit of truth: A critical examination of meta-analyses of cognitive behavior therapy.* Psychotherapy Research, 2017. **27**(1): p. 14-32.
62. Clark, G.I. and S.J. Egan, *The Socratic method in cognitive behavioural therapy: A narrative review.* Cognitive Therapy and Research, 2015. **39**(6): p. 863-879.
63. Nelson, L., *The socratic method.* Thinking: The Journal of Philosophy for Children, 1980. **2**(2): p. 34-38.
64. Zare, P. and J. Mukundan, *The use of Socratic method as a teaching/learning tool to develop students' critical thinking: A review of literature.* Language in India, 2015. **15**(6): p. 256-265.
65. Martina Heinemann, M.V.S.a.M.H., *Is it real? Dealing with an insecure perception of a pet robot in dementia care,* in. *Proceedings of New Friends 2015 – The 1st international conference on Social Robots in Therapy and Education,.* 2015: Almere, The Netherlands.
66. Soler, M.V. and R. Bemelmans, *Cuddling With New Friends.* 2014, Almere: Windesheim Flevoland.
67. Smits, C., et al., *Towards practical guidelines and recommendations for using robotics pets with dementia patients.* Can Int J Soc Sci Educ, 2015. **3**: p. 656-70.

68. Conrad, C.A., *Basic of Ethics*, in *Business Ethics-A Philosophical and Behavioral Approach*. 2018, Springer. p. 1-10.

69. Singer, P., *Practical ethics*. 2011: Cambridge university press.

70. Borenstein, J. and R. Arkin, *Robotic nudges: the ethics of engineering a more socially just human being*. Science and engineering ethics, 2016. **22**(1): p. 31-46.

71. Luxton, D.D., *Recommendations for the ethical use and design of artificial intelligent care providers*. Artificial intelligence in medicine, 2014. **62**(1): p. 1-10.

72. Riek, L. and D. Howard, *A code of ethics for the human-robot interaction profession*, in *We Robot*. 2014: Coral Gables, Florida, USA.

73. Veruggio, G. and F. Operto, *Roboethics: Social and ethical implications of robotics*, in *Springer handbook of robotics*. 2008, Springer. p. 1499-1524.

74. Katz, R., *Tele-care robot for assisting independent senior citizens who live at home*. 2015.

75. Rumeau, P., et al. *Home deployment of a doubt removal telecare service for cognitively impaired elderly people: a field deployment*. in *Cognitive Infocommunications (CogInfoCom), 2012 IEEE 3rd International Conference on*. 2012. IEEE.

76. Lovo Grona, S., et al., *Case Report: Using a Remote Presence Robot to Improve Access to Physical Therapy for People with Chronic Back Disorders in an Underserved Community*. Physiotherapy Canada, 2017. **69**(1): p. 14-19.

77. Arvan, M., *Humans and Hosts in Westworld*. Westworld and Philosophy: If You Go Looking for the Truth, Get the Whole Thing, 2018: p. 26-37.

78. González, L.C., *Turing's Dream and Searle's Nightmare in Westworld.* Westworld and Philosophy: If You Go Looking for the Truth, Get the Whole Thing, 2018: p. 71-78.

79. Hirvonen, O., *Westworld.* Westworld and Philosophy: If You Go Looking for the Truth, Get the Whole Thing, 2018: p. 61-70.

80. Irwin, W., *Westworld and Philosophy.* 2018: John Wiley & Sons.

81. Lyons, S., *Crossing the Uncanny Valley.* Westworld and Philosophy: If You Go Looking for the Truth, Get the Whole Thing, 2018: p. 39-49.

82. Trapero-Llobera, P., *The Observer (s) System and the Semiotics of Virtuality in Westworld's Characters.* Westworld and Philosophy: If You Go Looking for the Truth, Get the Whole Thing, 2018: p. 162-172.

83. Kind, A., *Blade Runner 2049.* The Philosophers' Magazine, 2018(80): p. 108-110.

84. Bukatman, S., *Blade runner.* 2017: Bloomsbury Publishing.

85. Gruzinski, S., *La guerra de las imágenes. De Cristóbal Colón a Blade Runner (1942-2019).* 2012: Fondo de Cultura Económica.

86. Landsberg, A., *Prosthetic memory: Total recall and blade runner.* Body & Society, 1995. 1(3-4): p. 175-189.

87. Kerman, J.B., *Technology and politics in the Blade Runner dystopia.* Retrofitting Blade runner: Issues in Ridley Scott's Blade runner and Philip K. Dick's Do androids dream of electric sheep, 1991: p. 16-24.

88. Kerman, J., *Retrofitting Blade runner: issues in Ridley Scott's Blade runner and Philip K. Dick's Do

androids dream of electric sheep? 1991: Popular Press.

89. Bruno, G., *Ramble City: Postmodernism and" Blade Runner".* October, 1987. **41**: p. 61-74.

90. Desser, D., *Blade Runner: Science Fiction & Transcendence.* Literature/Film Quarterly, 1985. **13**(3): p. 172.

91. Jackson, F., *Epiphenomenal qualia.* The Philosophical Quarterly (1950-), 1982. **32**(127): p. 127-136.

92. Jackson, F., *What Mary didn't know.* The Journal of Philosophy, 1986. **83**(5): p. 291-295.

93. Day, D.D., *Monty Python and the Holy Grail: madness with a definite method.* Cinema Arthuriana, 2002: p. 127-35.

94. Hawthorne, J. and G. Uzquiano, *How many angels can dance on the point of a needle? Transcendental theology meets modal metaphysics.* Mind, 2011. **120**(477): p. 53-81.

95. Harrison, P., *Angels on Pinheads and Needles' Points.* Notes and Queries, 2016. **63**(1): p. 45-47.

96. Nijholt, A. *Embodied agents: A new impetus to humor research.* in *The April Fools Day Workshop on Computational Humour.* 2002. In: Proc. Twente Workshop on Language Technology.

97. Dybala, P., et al. *Humoroids: conversational agents that induce positive emotions with humor.* in *Proceedings of The 8th International Conference on Autonomous Agents and Multiagent Systems-Volume 2.* 2009. International Foundation for Autonomous Agents and Multiagent Systems.

98. Mihalcea, R. and C. Strapparava, *Learning to laugh (automatically): Computational models for humor*

recognition. Computational Intelligence, 2006. **22**(2): p. 126-142.

99. Baird, H.S., A.L. Coates, and R.J. Fateman, *Pessimalprint: a reverse turing test.* International Journal on Document Analysis and Recognition, 2003. **5**(2-3): p. 158-163.

100. Moor, J.H., *An analysis of the Turing test.* Philosophical Studies, 1976. **30**(4): p. 249-257.

101. Saygin, A.P., I. Cicekli, and V. Akman, *Turing test: 50 years later.* Minds and machines, 2000. **10**(4): p. 463-518.

102. Shieber, S.M., *The Turing test: Verbal behavior as the hallmark of intelligence.* 2004: Mit Press.

103. Turing, A.M., *Computing machinery and intelligence*, in *Parsing the Turing Test.* 2009, Springer. p. 23-65.

104. Heerink, M., *Zolang je robot maar van je houdt – Hoe en waarom onze wereld verandert door de onstuitbare opkomst van de robots* 2013, Schiedam: Scriptum.

105. Searle, J., *The Chinese Room.* 1999.

106. Searle, J.R., *Minds, brains, and programs.* Behavioral and brain sciences, 1980. **3**(3): p. 417-424.

107. Preston, J. and M.J. Bishop, *Views into the Chinese room: New essays on Searle and artificial intelligence.* 2002: OUP.

108. Searle, J.R., D.C. Dennett, and D.J. Chalmers, *The mystery of consciousness.* 1997: New York Review of Books.

109. Albo-Canals, J., et al., *A Pilot Study of the KIBO Robot in Children with Severe ASD.* International Journal of Social Robotics, 2018: p. 1-13.

110. Albo-Canals, J., et al. *Comparing two LEGO Robotics-based interventions for social skills training with children with ASD*. in *RO-MAN, 2013 IEEE*. 2013. IEEE.

111. Barua, R., S. Sramon, and M. Heerink, *Empathy, compassion and social robots: an approach from Buddhist philosophy*, in *New Friends 2015*. 2015: Almere.

112. Rosenthal-von der Pütten, A.M., et al., *An experimental study on emotional reactions towards a robot*. International Journal of Social Robotics, 2013. **5**(1): p. 17-34.

113. Rosenthal-Von Der Pütten, A.M., et al., *Investigations on empathy towards humans and robots using fMRI*. Computers in Human Behavior, 2014. **33**: p. 201-212.

114. Darling, K., P. Nandy, and C. Breazeal. *Empathic concern and the effect of stories in human-robot interaction*. in *Robot and Human Interactive Communication (RO-MAN), 2015 24th IEEE International Symposium on*. 2015. IEEE.

115. Asada, M., *Towards artificial empathy*. International Journal of Social Robotics, 2015. **7**(1): p. 19-33.

116. Paiva, A., et al., *Empathy in Virtual Agents and Robots: A Survey*. ACM Transactions on Interactive Intelligent Systems (TiiS), 2017. **7**(3): p. 11.

117. Stipek, D.J., *Motivation to learn: Integrating theory and practice*. 2002: Allyn and Bacon Boston, MA.

118. Saskia van Oenen, H.v.K.a.M.H., *Social Robotics In Education Involving ASD Children: A Collaborative Design Project*, in *Proceedings New Friends 2016 - The 2nd international conference on*

Social Robots in Therapy and Education. 2016: Barcelona, Spain.

119. Russell, A.M.T. and S.T. Fiske, *It's all relative: Competition and status drive interpersonal perception.* European Journal of Social Psychology, 2008. **38**(7): p. 1193-1201.

120. Fisher, R.M., *Love and fear.* A CSIIE Yellow Paper, DIFS-6. Carbondale, IL: Center for Spiritual Inquiry and Integral Education, 2012.

121. Fisher, H.E., et al., *Defining the brain systems of lust, romantic attraction, and attachment.* Archives of sexual behavior, 2002. **31**(5): p. 413-419.

122. Dutton, D.G. and A.P. Aron, *Some evidence for heightened sexual attraction under conditions of high anxiety.* Journal of personality and social psychology, 1974. **30**(4): p. 510.

123. Heerink, M., *Assessing acceptance of assistive social robots by aging adults.* 2010, Universiteit van Amsterdam [Host].

124. Heerink, M., et al., *Assessing acceptance of assistive social agent technology by older adults: the almere model.* International journal of social robotics, 2010. **2**(4): p. 361-375.

125. Can, W.S.R. and S.D.J. Seibt, *Social robotics, elderly care, and human dignity: a recognition-theoretical approach.* What Social Robots Can and Should Do: Proceedings of Robophilosophy 2016/TRANSOR 2016, 2016. **290**: p. 155.

126. Wikipedia. *Soul.* 2018 [cited 2018 09-10-2018]; Available from: https://en.wikipedia.org/wiki/Soul.

127. Sillar, B., *The social agency of things? Animism and materiality in the Andes.* Cambridge Archaeological Journal, 2009. **19**(3): p. 367-377.

128. Harvey, G., *Animism: Respecting the living world.* 2005: Wakefield Press.
129. Willerslev, R., *Soul hunters.* Hunting, animism, and personhood among the Siberian Yukaghirs, 2007.
130. Swinburne, R., *The evolution of the soul.* 1986.
131. Spezzano, C. and G.J. Gargiulo, *Soul on the couch: Spirituality, religion, and morality in contemporary psychoanalysis.* Vol. 7. 2013: Routledge.
132. Hofstadter, D.R. and D.C. Dennett, *The Mind\'s I: Fantasies and Reflections on Self & Soul.* 2006.
133. Lindenfors, P., *Your Psychological Self*, in *For Whose Benefit?* 2017, Springer. p. 27-35.
134. Vary, A., *Spiritual Nature of Consciousness.* Journal of Consciousness Exploration & Research, 2017. **8**(2).
135. Burgin, M., *Ideas of Plato in the context of contemporary science and mathematics.* Athens Journal of Humanities and Arts, 2017. **4**: p. 161-182.
136. Pfattheicher, S., J. Keller, and G. Knezevic, *Destroying things for pleasure: On the relation of sadism and vandalism.* Personality and Individual Differences, 2018.
137. Mafimisebi, O.P. and S. Thorne, *Vandalism-militancy relationship: the influence of risk perception and moral disengagement.* International Journal of Mass Emergencies and Disasters, 2017. **35**(3): p. 191-223.
138. Bhati, A. and P. Pearce, *Tourist behaviour, vandalism and stakeholder responses.* Visitor management in tourism destinations, 2017: p. 102-116.
139. Bhati, A. and P. Pearce, *Tourist attractions in Bangkok and Singapore; linking vandalism and*

setting characteristics*. Tourism Management, 2017.
63: p. 15-30.

140. Muller, G.B., *Vestigial organs and structures.*, in *Encyclopedia of evolution,*, M. Pagel, Editor. 2002, Oxford University Press: New York. p. 1131-1133.

141. Russell, B., *Robots: The 500-year Quest to Make Machines Human.* 2017: Scala Arts Publishers, Inc.

142. Mahum, R., et al., *A review on humanoid robots.* INTERNATIONAL JOURNAL OF ADVANCED AND APPLIED SCIENCES, 2017. **4**(2): p. 83-90.

143. Finn, R.V., *"More Human Than Human": Lacan's Mirror Stage Theory and Posthumanism in Philip K. Dick's Do Androids Dream of Electric Sheep?* 2018.

144. Matsui, D., et al., *Generating natural motion in an android by mapping human motion.* 2018: Springer.

145. Sell, A., J. Tooby, and L. Cosmides, *Formidability and the logic of human anger.* Proceedings of the National Academy of Sciences, 2009. **106**(35): p. 15073-15078.

146. Prahl, A., *Mortal Versus Machine: Investigating Interpersonal Advice and Automated Advice.* 2018: The University of Wisconsin-Madison.

147. Ellis, A., *Anger: How to live with and without it.* 2017: Citadel.

148. Enright, R.D. and R.P. Fitzgibbons, *Forgiveness therapy: An empirical guide for resolving anger and restoring hope.* 2015: American Psychological Association.

149. Bandura, A., *Social cognitive theory of moral thought and action*, in *Handbook of moral behavior and development.* 2014, Psychology Press. p. 69-128.

150. Kohlberg, L., *Development of moral character and moral ideology*. Review of child development research, 1964. **1**: p. 381-431.
151. Killen, M. and J.G. Smetana, *Handbook of moral development*. 2013: Psychology Press.
152. Kohlberg, L. and R. Kramer, *Continuities and discontinuities in childhood and adult moral development*. Human development, 1969. **12**(2): p. 93-120.
153. Asimov, I., *I, robot*. Vol. 1. 2004: Spectra.
154. Foot, P., *The Problem of Abortion and the Doctrine of the Double Effect*, in *Virtues and Vices*, B. Blackwell, Editor. 1978: Oxford.
155. de Ruiter, H.-P., *The safety paradox*. Creative nursing, 2006. **12**(3): p. 4.
156. Engwicht, D., *Solving the safety paradox--when making things less safe makes them more safe*. Injury prevention, 2012. **18**(Suppl. 1): p. A1-A1.
157. Matsuzaki, H. and G. Lindemann, *The autonomy-safety-paradox of service robotics in Europe and Japan: a comparative analysis*. AI & society, 2016. **31**(4): p. 501-517.
158. Dennett, D.C., *Kinds of minds: Toward an understanding of consciousness*. 2008: Basic Books.
159. Chalmers, D.J., *The character of consciousness*. 2010: Oxford University Press.
160. Heerink, M. *How elderly users of a socially interactive robot experience adaptiveness, adaptability and user control*. in *12th IEEE 2011 CINTI Conf.* 2011.
161. Cabrera, A., W.C. Collins, and J.F. Salgado, *Determinants of individual engagement in knowledge sharing*. The International Journal of

Human Resource Management, 2006. **17**(2): p. 245-264.

162. Lin, H.-F., *Effects of extrinsic and intrinsic motivation on employee knowledge sharing intentions.* Journal of information science, 2007. **33**(2): p. 135-149.

163. Matzler, K., et al., *Personality traits and knowledge sharing.* Journal of economic psychology, 2008. **29**(3): p. 301-313.

164. Wang, S. and R.A. Noe, *Knowledge sharing: A review and directions for future research.* Human resource management review, 2010. **20**(2): p. 115-131.

165. Blum, P.R., *Robots, Slaves, and the Paradox of the Human Condition in Isaac Asimov's Robot Stories.* Roczniki Kulturoznawcze, 2017. **7**(3): p. 5-24.

166. Kim, M.-S., *Robot as the "mechanical other": transcending karmic dilemma.* AI & SOCIETY, 2018: p. 1-10.

167. Latham, R., *Science Fiction Criticism: An Anthology of Essential Writings.* 2017: Bloomsbury Publishing.

Films and Television series

- *2001—A Space Oddyssee*—Stanley Kubrick, 1986

  Features HAL, actually a computer and not a robot, but it's an artificial intelligence with definitely some dominant traits.

- *A.I. Artificial Intelligence*—Steven Spielberg, 2001

  A bit of a long story, but filmed brilliantly. Pay particular attention to the amazing Gigolo Joe.

- *Astro Boy*—David Bowers, 2009

  Astro Boy is a classic Japanese robot. Illustrative of how much they love robots in Japan.

- *Autofac (from the Electric Dreams series)*—Peter Horton, 2018

  Nice series too, based on stories by Philip K. Dick.

- *Bicentennial Man*—Chris Columbus, 1999

With a robot that gradually turns into Robin Williams with the unforgettable words 'I would rather die a man, than live for all eternity as a machine'.

- *Big Hero 6*—Don Hall, Chris Williams, 2014

  Notice how much expressiveness a robot can have without a facial expression. And pay attention to the dangerous swarm.

- *Blade Runner*—Ridley Scott, 1982

  My #1. With the unforgettable monologue. 'All those moments will be lost, like tears in the rain'.

- *Blade Runner 2049*—Denis Villeneuve, 2017

  And what if robots can reproduce one day ...?

- *Cherry 2000—Steve De Jarnatt, 1987*

  A kind of cult classic with a special erotic human-robot scene, including dishwashing gloves.

- *Dark Matter*—Joseph Mallozzi, Paul Mullie, 2015- (TV series)

  With an Android developing positively.

- *Ex Machina*—Alex Garland, 2014

  Strongly played and illustrative of what can happen in theory with a self-learning intelligent system with awareness.

- *Ghost in the Shell*—Mamoru Oshii, 1995

Wonderful animation and the story makes you think about human identity. In 2017 there was a non-animated version of Rupert Sanders, with Scarlett Johansson.

- *Her*—Spike Jonze, 2013

  Just like in 2001—A Space Oddyssee it is not a robot, but an 'operating system' with artificial intelligence. Illustrates the possibilities and limitations of human-system relations.

- *Humans*—Kudos Film and Television et al., 2015- (TV series)

  Particularly interesting because of the difference between android robots with and without consciousness.

- *I, Robot*—Alex Proyas, 2004

  The robots are not the problem, it is about the intelligent system that controls them. Fortunately, Will Smith saves us. Based on Isaac Asimov's work.

- *Lost in Space*—Irwin Allen, 1965-1968 (TV series), Stephen Hopkins 1998 (film), Irwin Allen, Matt Sazama, Burk Sharpless 2018- (TV series)

  The robot is good or bad, depending on who controls it.

- *Robocop*—Paul Verhoeven, 1987 and José Padilha, 2014

The hybrid identity of the robot is particularly interesting: it is built out of a human being and occasionally we see the remains of its human part. Somewhat comparable to Ghost in the Shell.

- *Robot & Frank*—Jake Schreier, 2012

  Demonstrates very accurately how a human-robot relationship can develop.

- *Screamers*—Christian Duguay, 1995

  Shows what can go wrong if you develop war robots that can also evolve. Strong point. Let's not do this, ever.

- *Short Circuit*—John Badham, 1986

  On how people can bond with a sometimes creaky robot. The robot in this film might well have been a model for Wall.e

- *Star Trek: First Contact*—Jonathan Frakes, 1996

  With 'the Borg', half human, half robot, who assimilate everyone they get hold of. Resistance is futile.

- *The Hitchhiker's Guide to the Galaxy*—Alan J.W. Bell, 1981 (TV series) and Garth Jennings, 2005 (film)

  With of course Marvin, the consistently depressed robot. Rightly so, by the way.

- *The Iron Giant*—Brad Bird, 1999

How to truly befriend a colossal, pretty scary robot.

- *The Terminator*—James Cameron, 1984

  Shows, also in the follow-up films, why it is crucial that a robot has some ethical awareness.

- *Wall.e*—Andrew Stanton, 2008

  Shows how an unsightly, animated robot can endear you. Especially thanks to the 'body language'.

- *Westworld*—Michael Crichton, 1973 (film), 2017 (TV series) Bad Robot et al.,

  Robots that you can shoot at will in a Western theme park, develop a form of consciousness and no longer accept their fate. The TV series goes a long way in this and addresses many themes mentioned above.

Index